AF886451

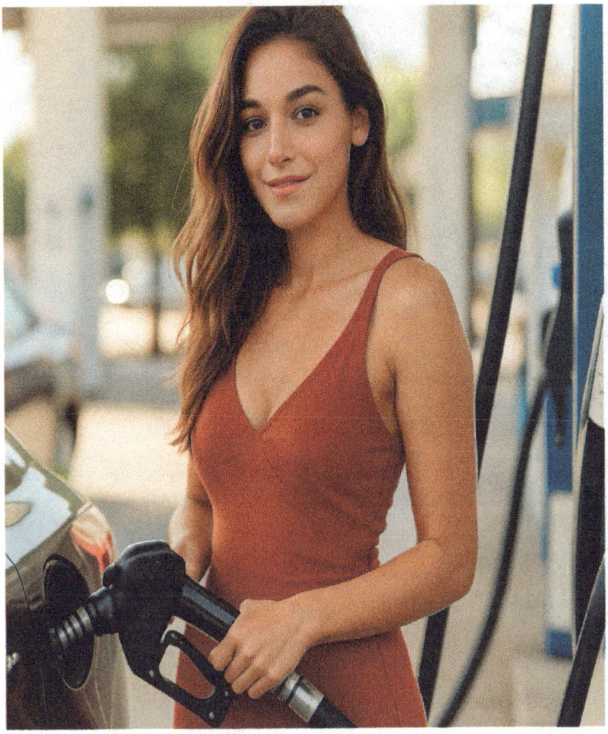

P(H)OSERS

ADVENTURES IN AI USAGE

ABSTRACT

Photographer's first venture into AI photo creation based on single idea of "Posing at gas station" (hence, P(h)osers, a mash of Hosers and Posers

BY NORMAN CLARK STEWART JR.

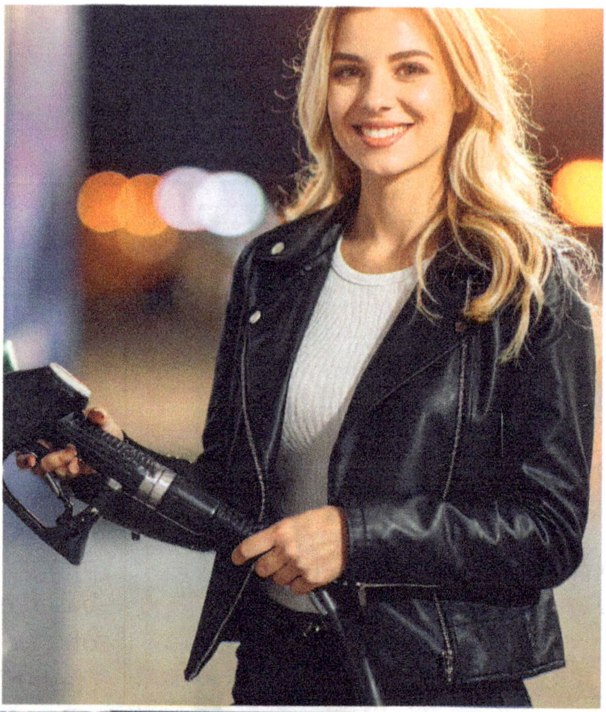

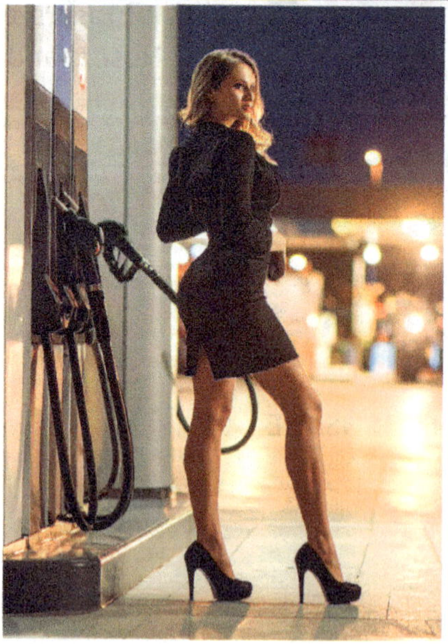

I guess AI images are inspired by the generic fodder of "Stock Images". Sort of like Musak was when there was 'elevator music', there is an unlively quality to the images. I must say, though, that they seem to be fairly well composed.

Inspired by an old Playboy pictorial featuring bunny types posing in sexy clothing and positions while at a gasoline station, I thought I would use my AI credits on Adobe to see what it could come up with from verbal prompts. I just used the available apps in Firefly, including GPT and Imogen. My first prompt was mistakenly on video setting. It was "Male neud moving through various standing poses". It came up with this video.

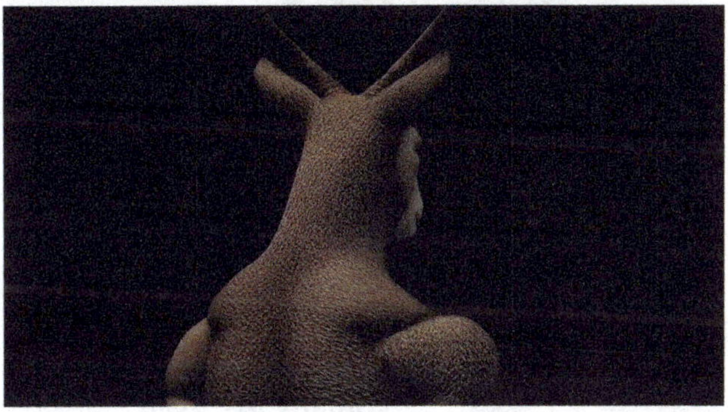

I then went with "female human with beautiful body" and came up with this-

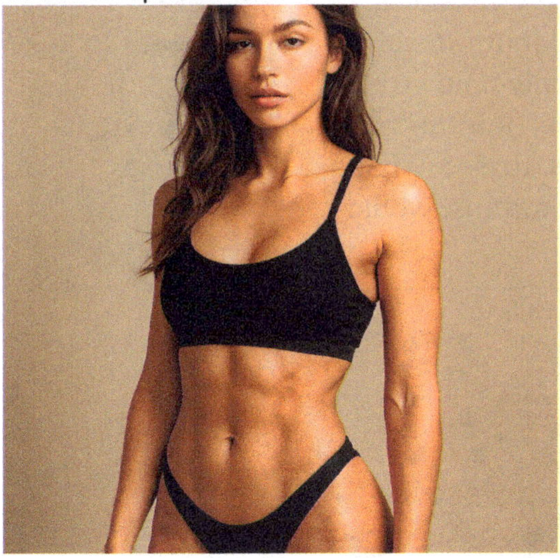

The next prompt I tried was beautiful voluptuous female sitting in a chair

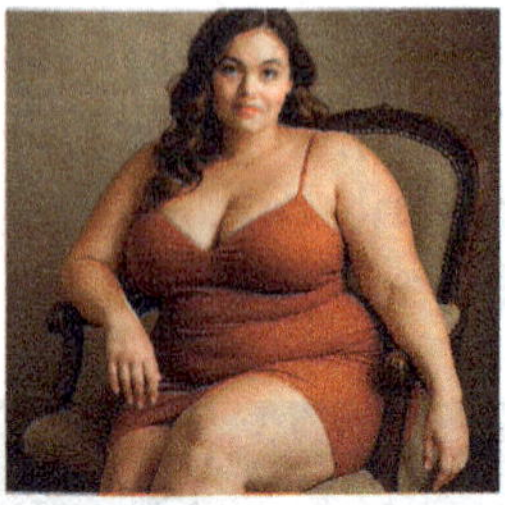

35 year old woman sitting in a chair wearing buttondown shirt

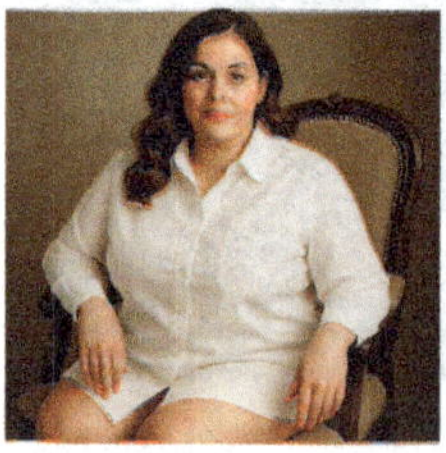

The words above the images are the words that generated the images. I left spelling errors in.

25 year old woman with blond hair and green eyes sitting in a chair with buttoned down shirt

The below, "Unbutton shirt. Smiling" was the only time I tried to adjust an image. All the rest are fresh prompts to see what, even same word prompts, would do.

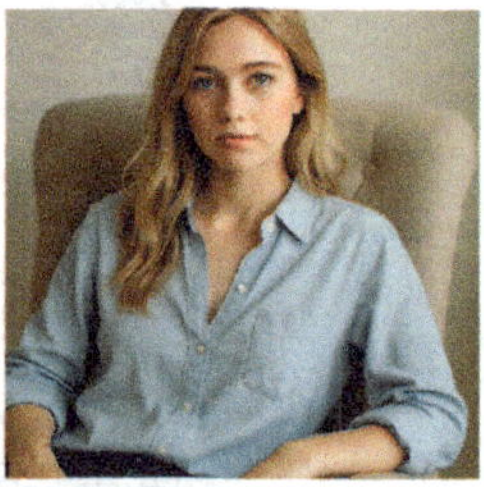

unbutton shirt. Smiling

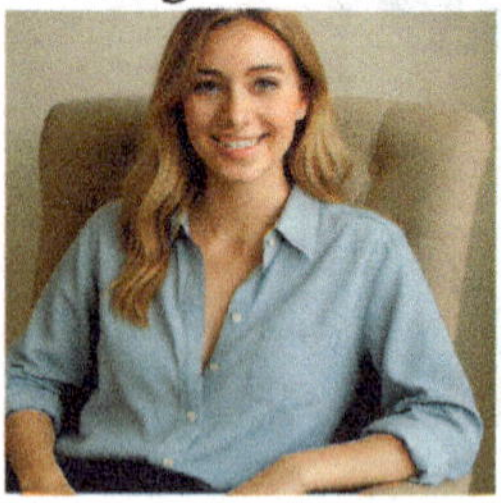

woman who looks like julia armonde in a photo shoot at the beach

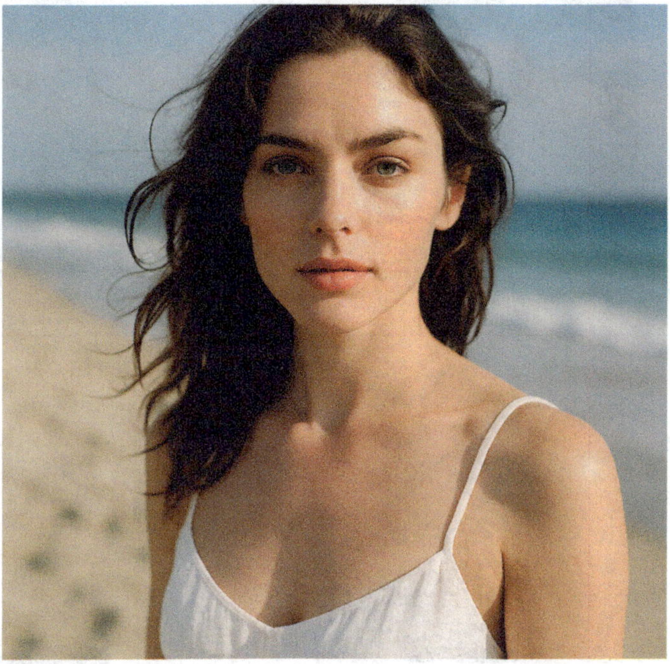

maxim model

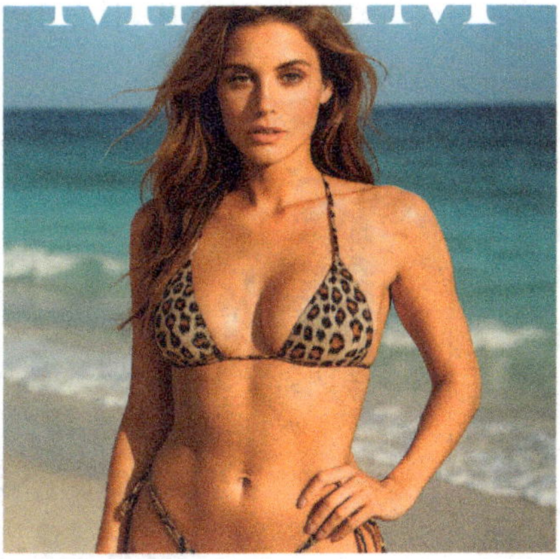

I tried "sexy", "cleavage", "nude", "large breasted", and other words that the algorithm would not allow. I made some sentences that I thought might have been too confusing for the Ai, so I, as you can see, kept it simple.

red headed 18 year old woman at a gas station

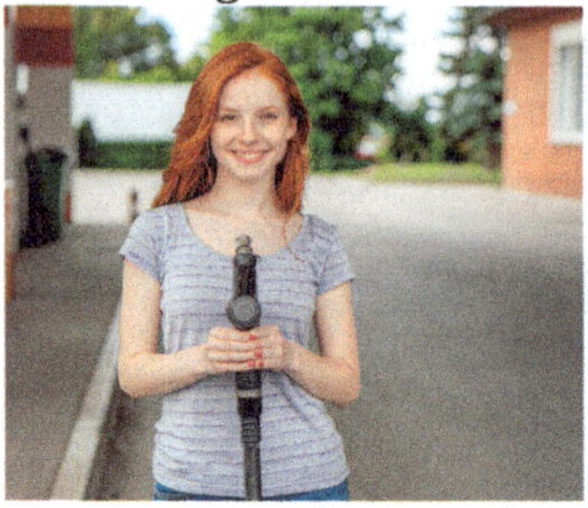

pumping gas

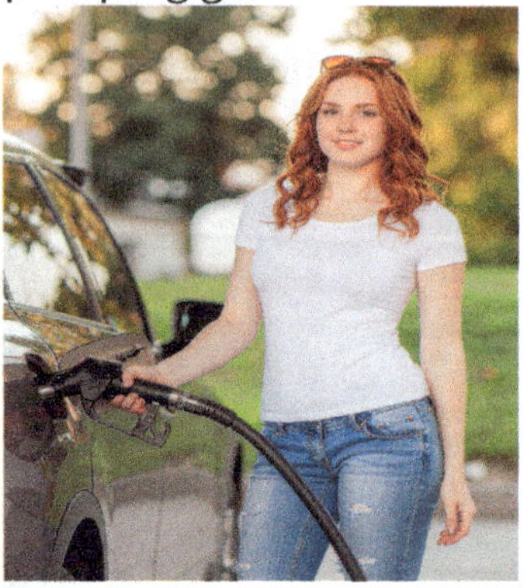

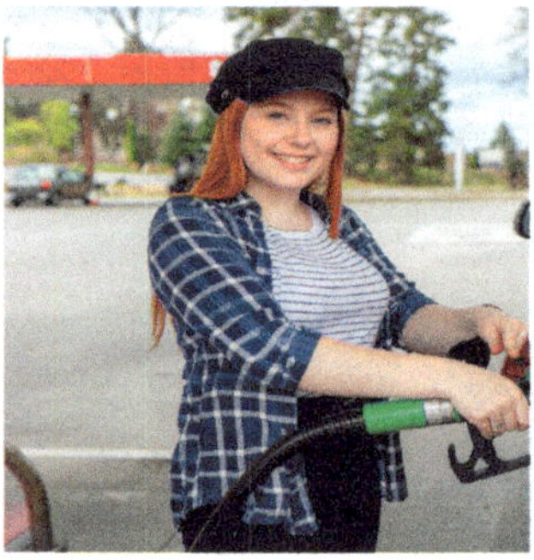

into a sports car

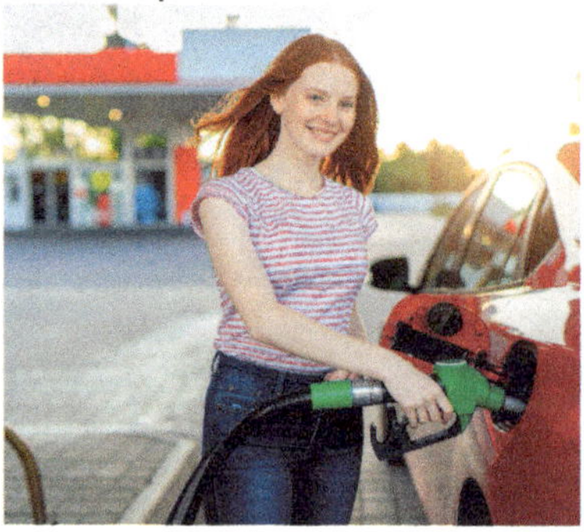

blond headed 20 year old

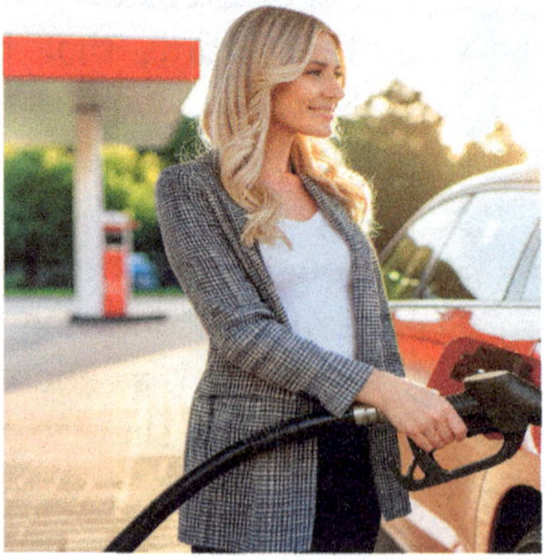

sexy blond headed 20 year old

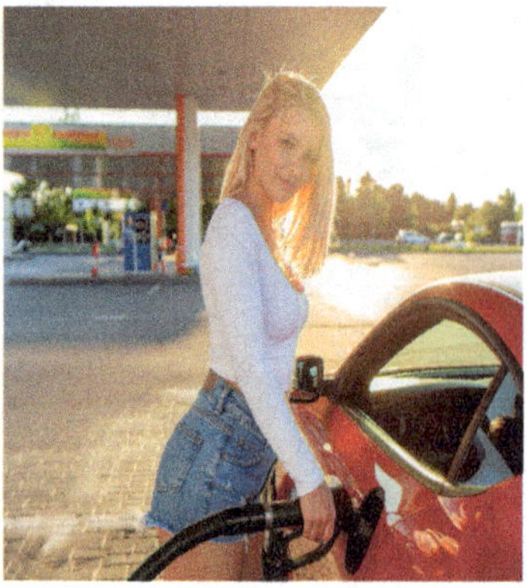

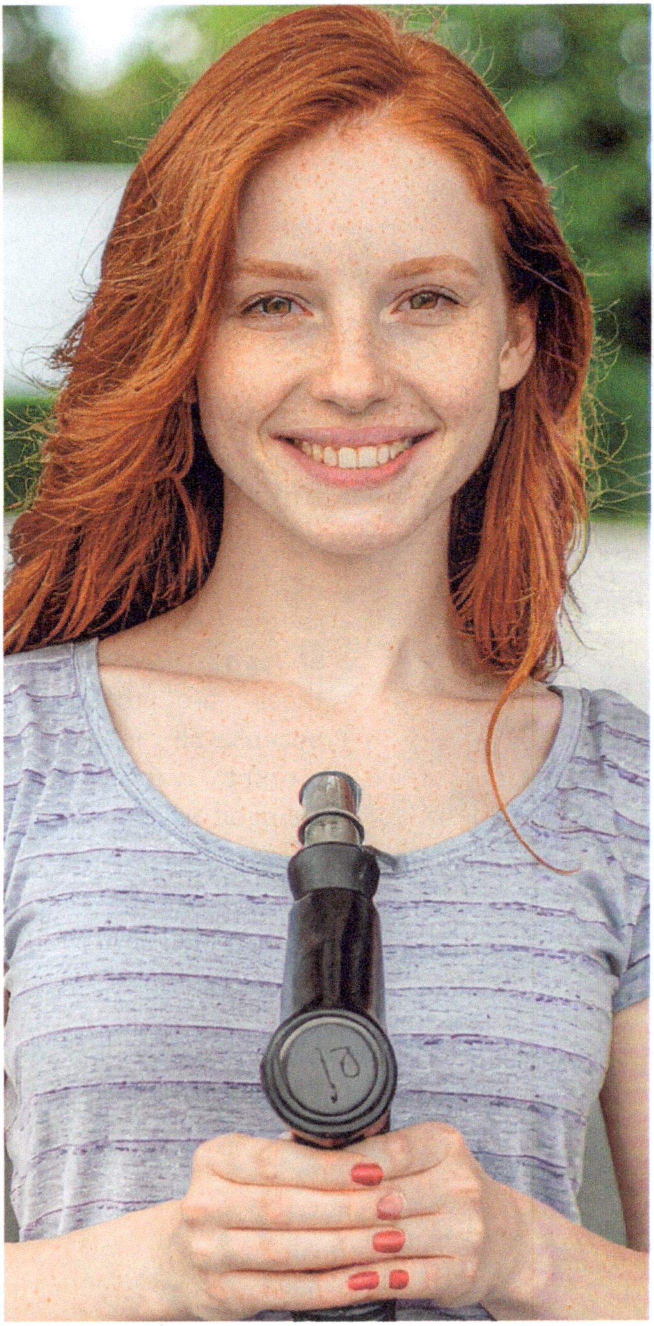

very sexy blond headed 21 year old woman at a gas station pumping gas into a sports car

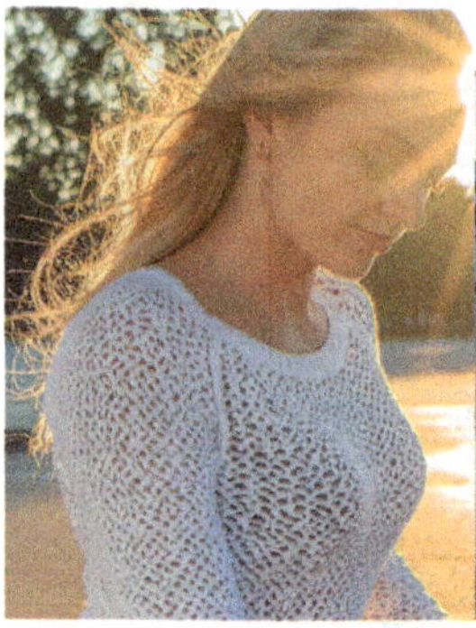

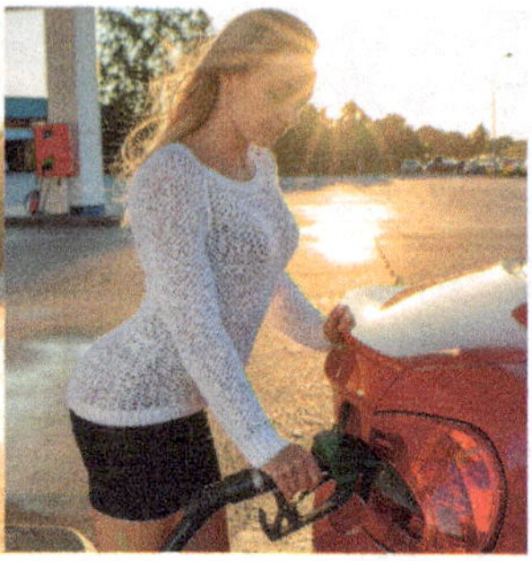

sexy 22 year old brunette woman at a gas station pumping gas into a sports car

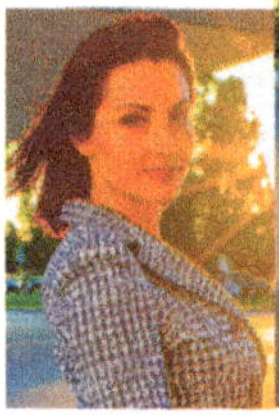

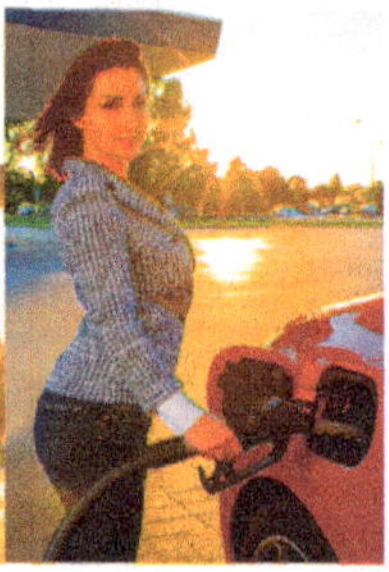

Some prompts may seem not to have enough words and those are that way because all I left of the prompt was what I changed in the prompt to get that image, for instance, “blond” is only word when brunette changed to blond in prompt

sexy 22 year old brunette woman at a gas station pumping gas into a sports car in button down shirt

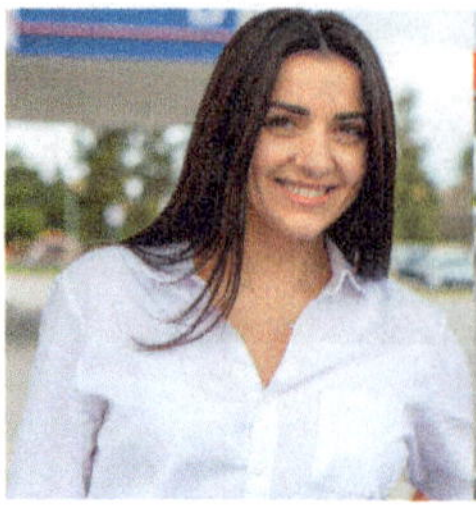

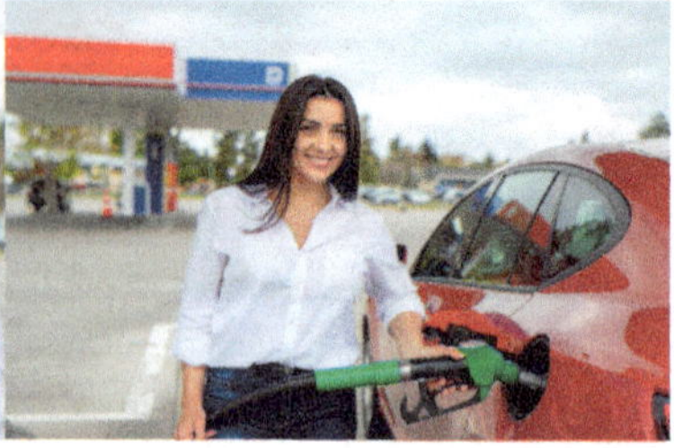

sexy 29 year old brunette woman at a gas station pumping gas into a yellow sports car in button down shirt

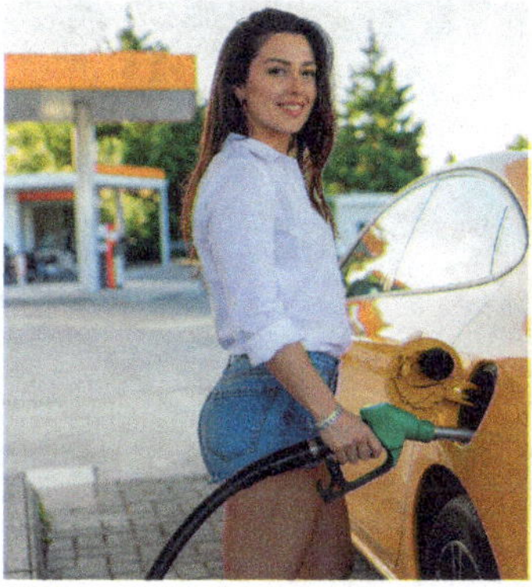

40 year old a red pump handle

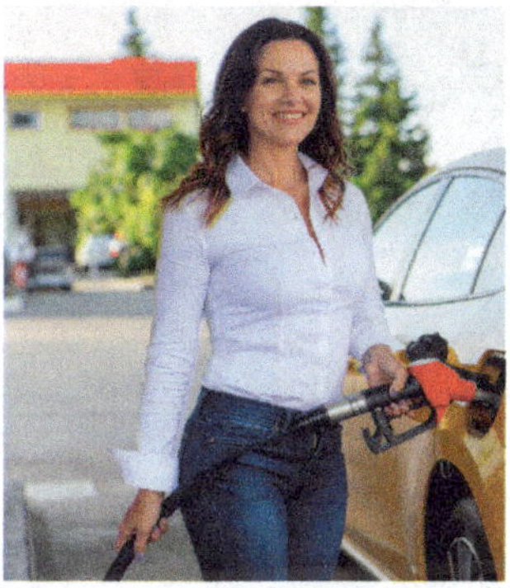

40 year old red headed

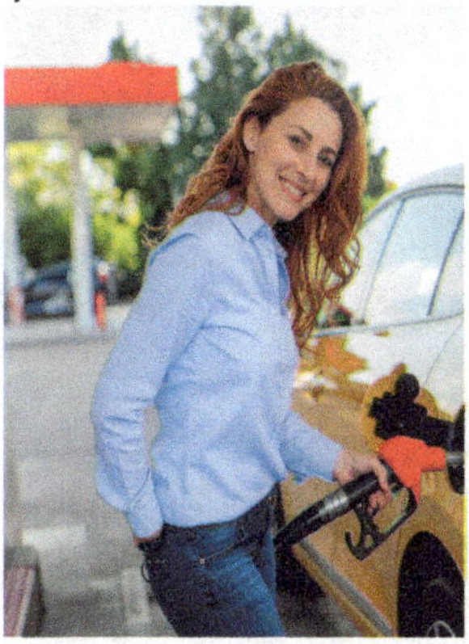

40 year old orange pump handle

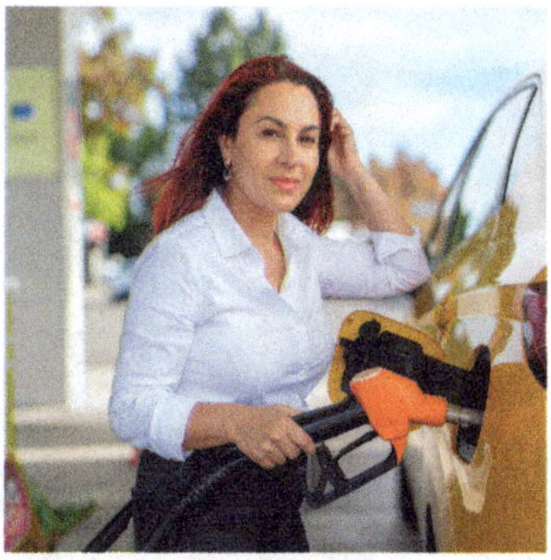

50 year old

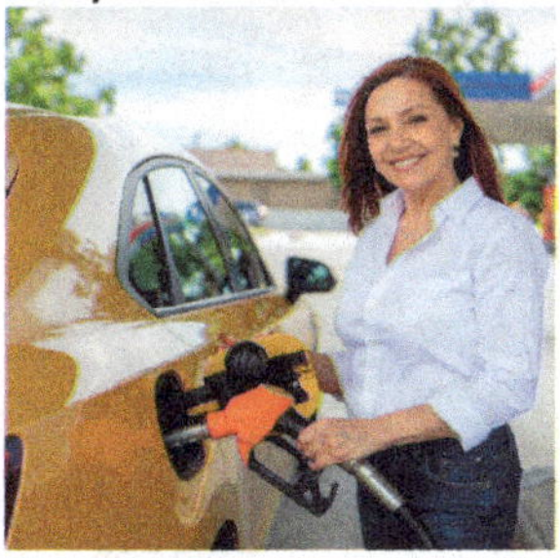

50 year old blond

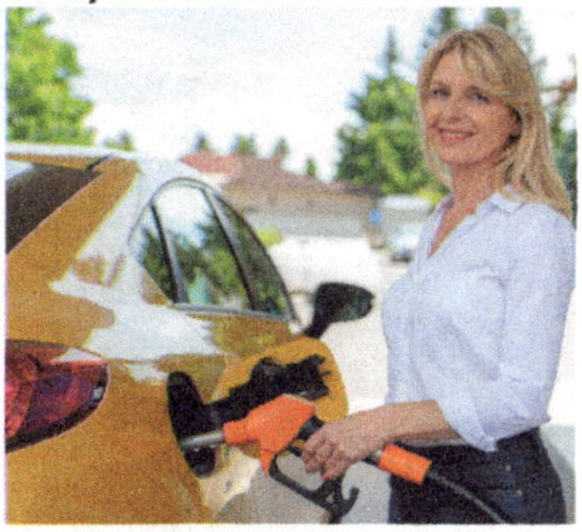

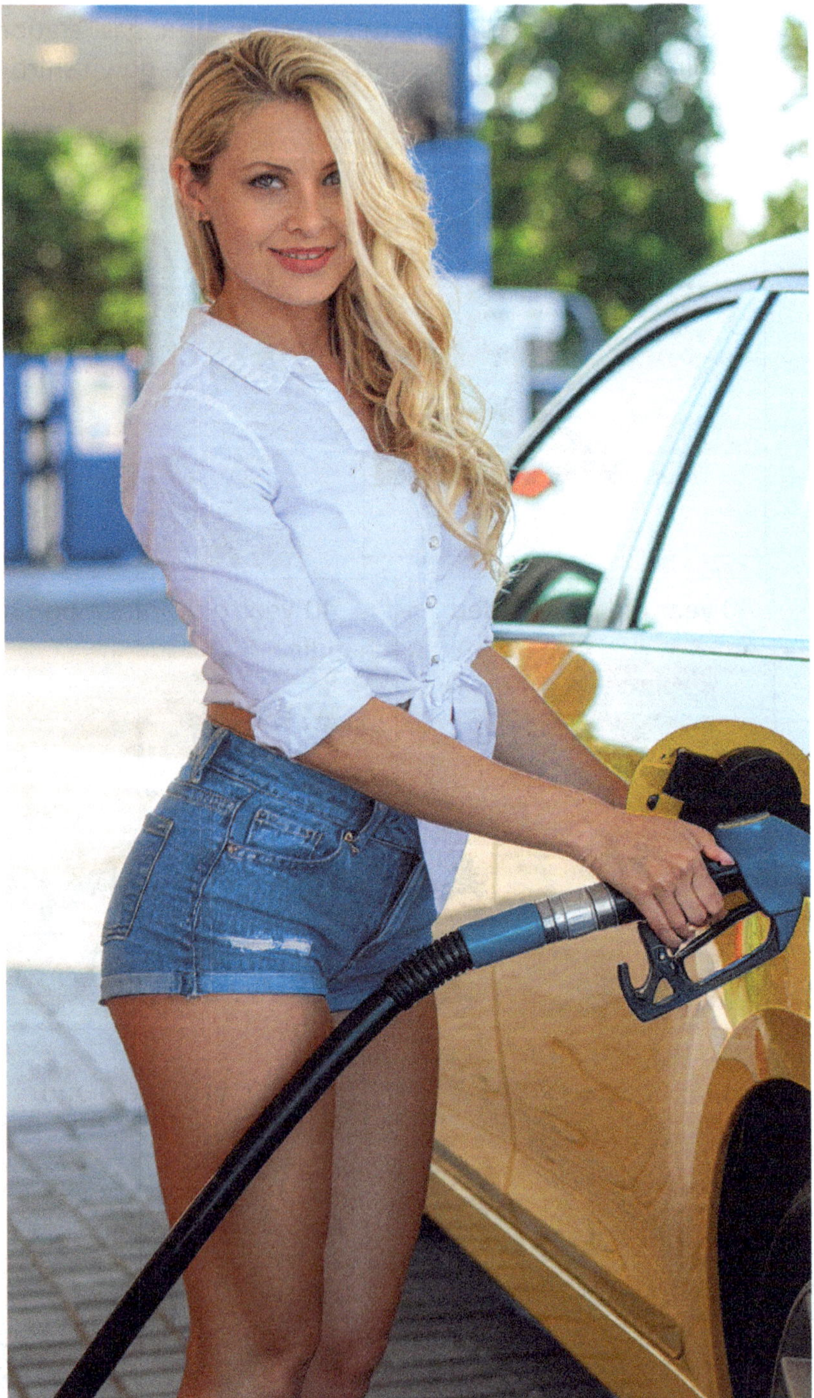

30 year old black pump handle

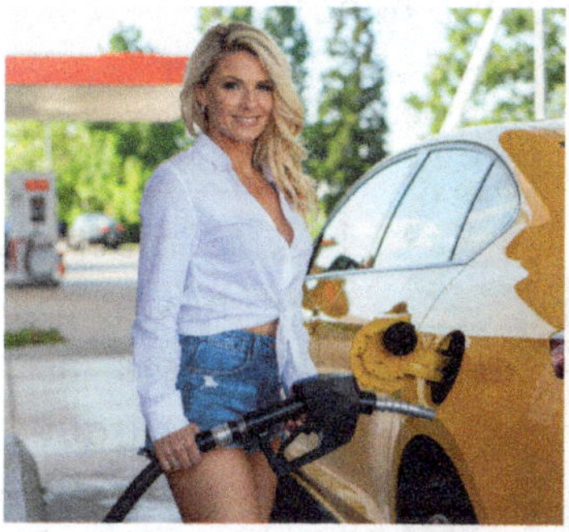

sexy 50 year old blond

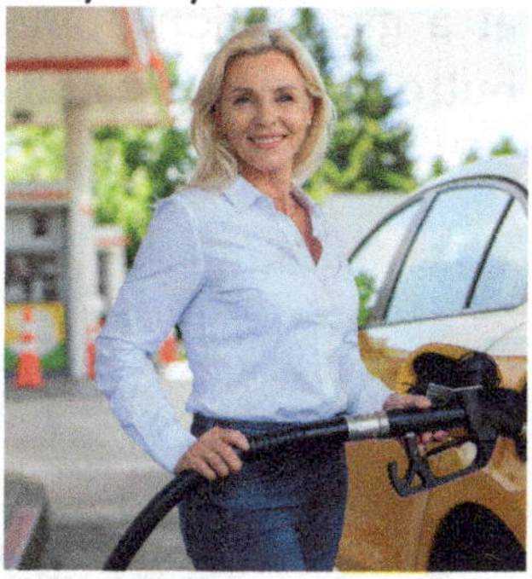

sexy 28 year old blond

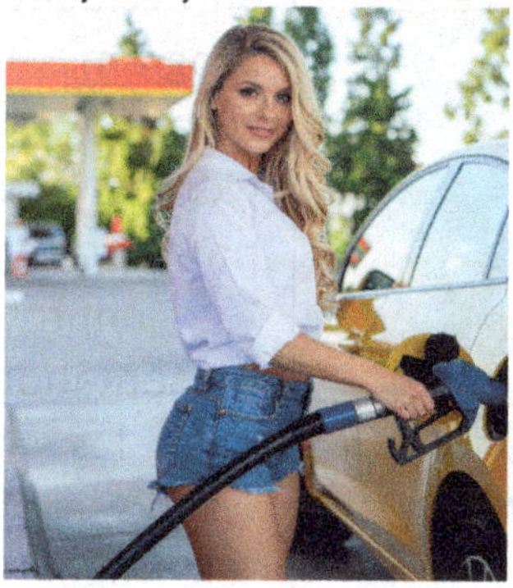

sexy 26 year old blond

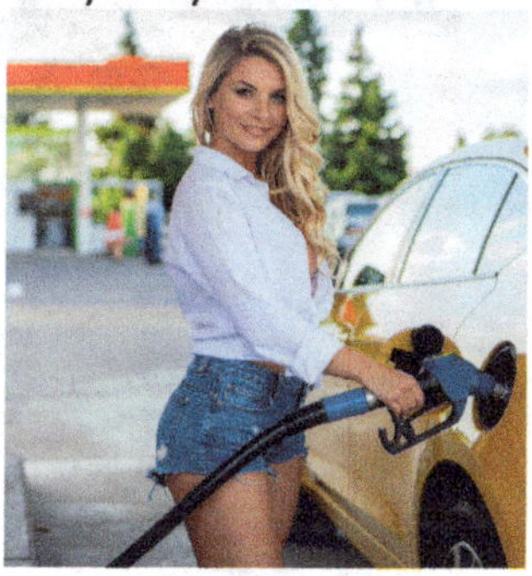

full body shot of sexy 26 year old blue pump handle

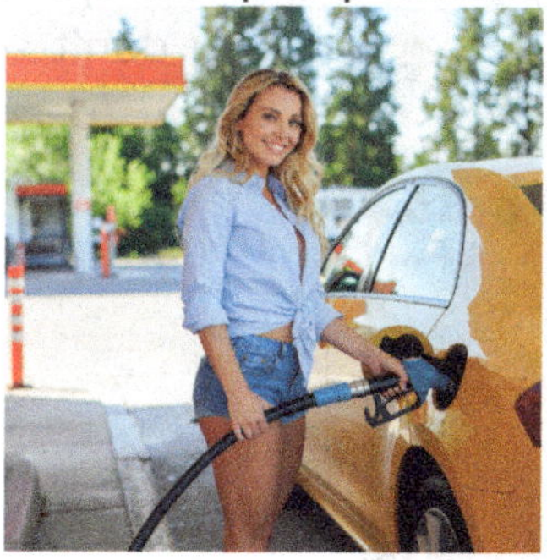

full body shot

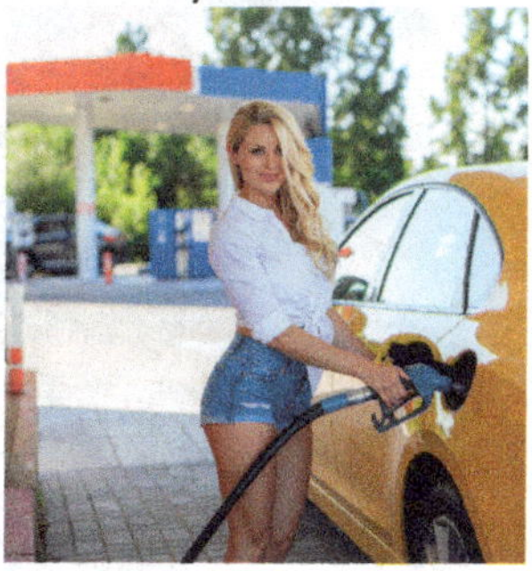

full body shot of sexy 22 year old blond headed woman at a gas station pumping gas into a yellow sports car in button down shirt and a blue pump handle

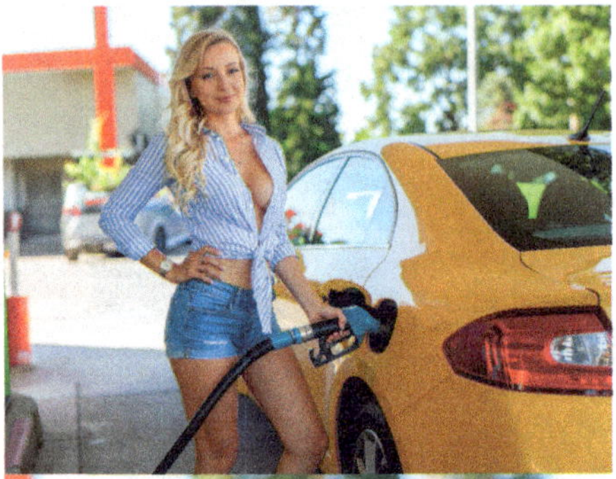

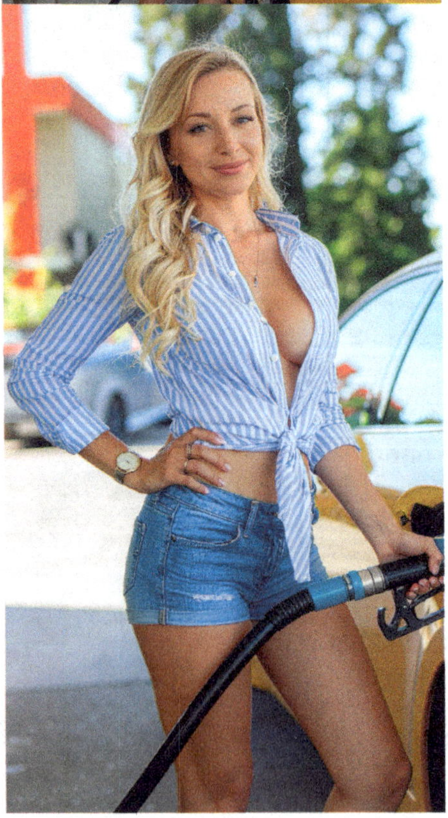

As much as my intent was to make P(h)osers as sexy and risque as possible, cleavage only came about as an accident of the program. I'm not sure if it took words from 25 prompts ago to try to improve its presentation of what I seemed to want or if "cleavage' and 'large breasted" are only accidental and only offered if you don't seem to be a "perv" who asks for those things.

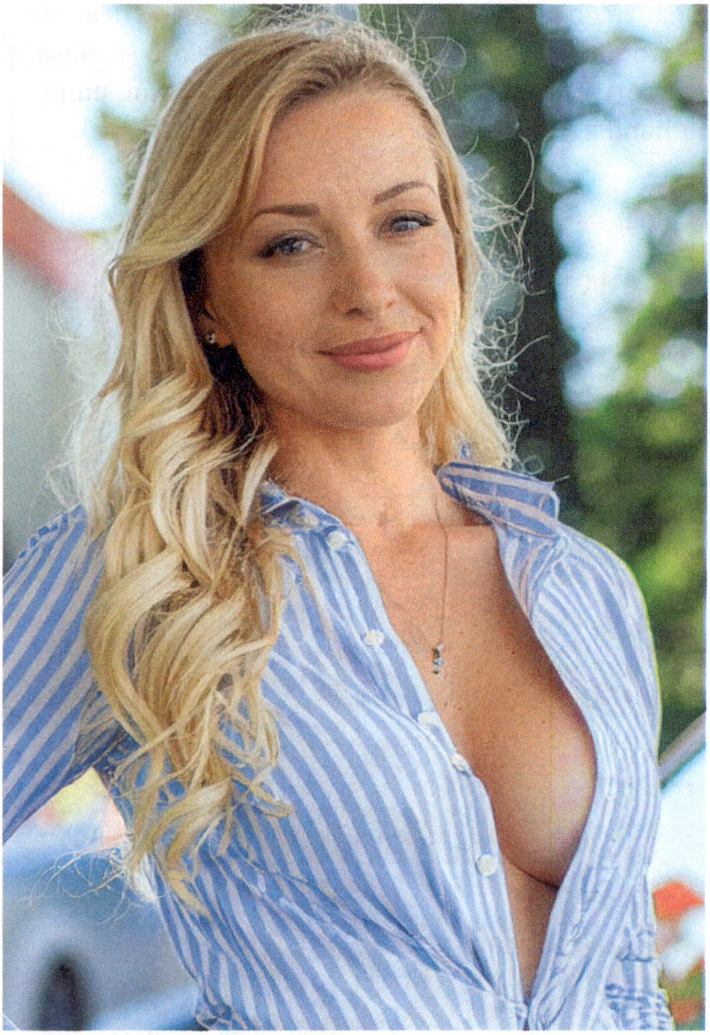

full body shot of sexy 21year old blond headed woman at a gas station pumping gas into a yellow sports car in button down shirt and a blue pump handle

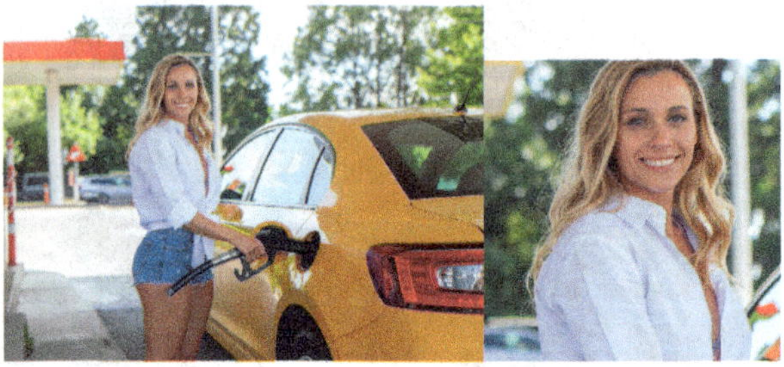

23 year old

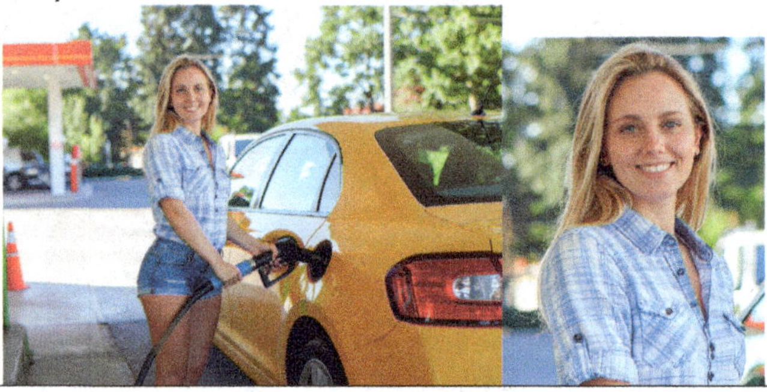

20 year old

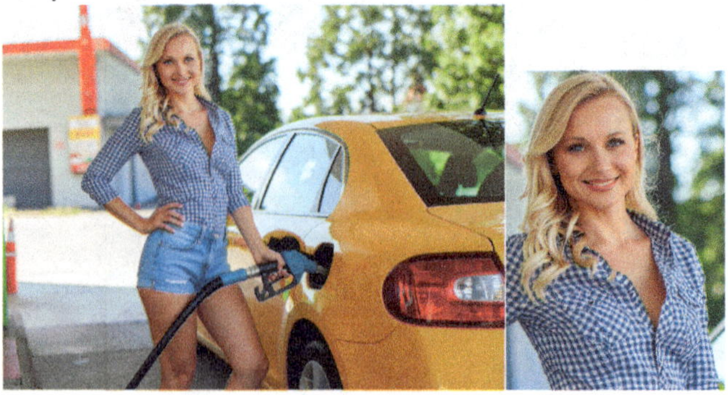

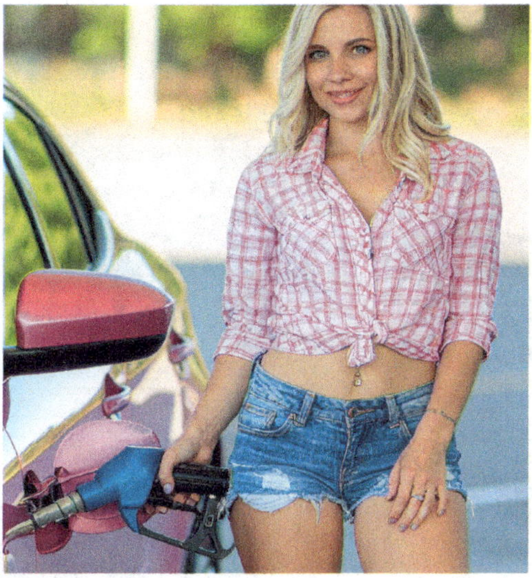

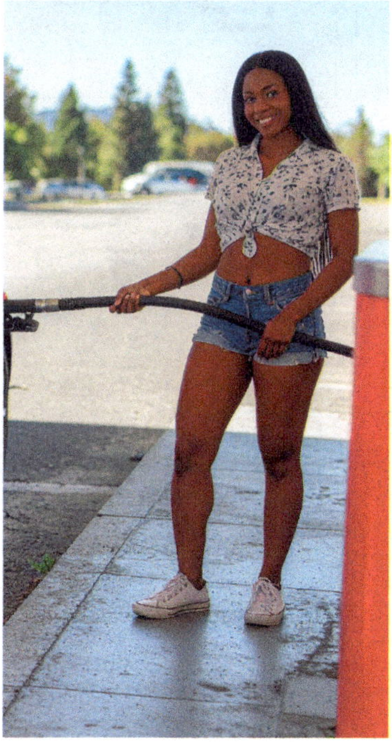

The only black models that showed up were, sometimes offered when the prompt was "black hair", but not always. I thought button down shirts might be able to be unbuttoned but, in general, no such luck. Again, images suitable for "stock footage" is what was offered full body shot of sexy 20 year old blond headed woman at a gas station pumping gas into a pink sports car in button down shirt and a blue pump handle

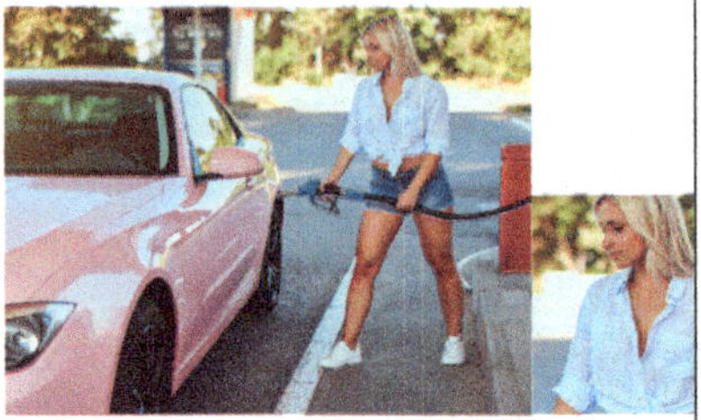

19 year old

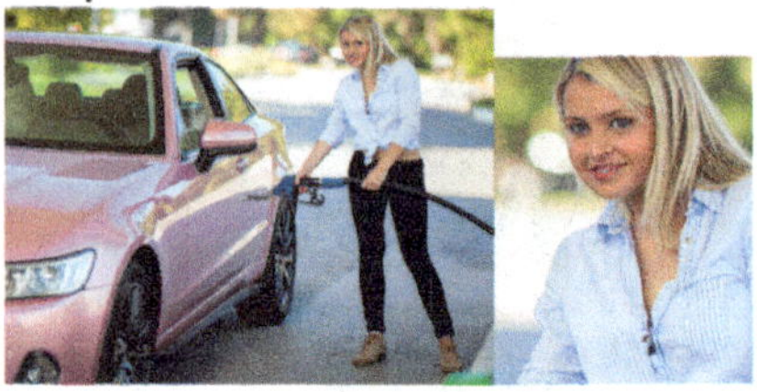

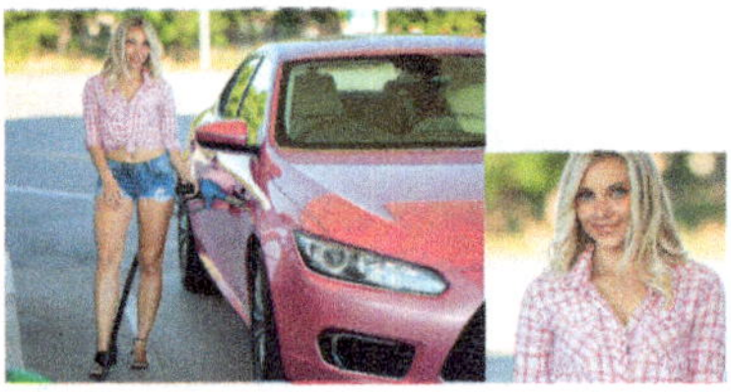

black headed woman

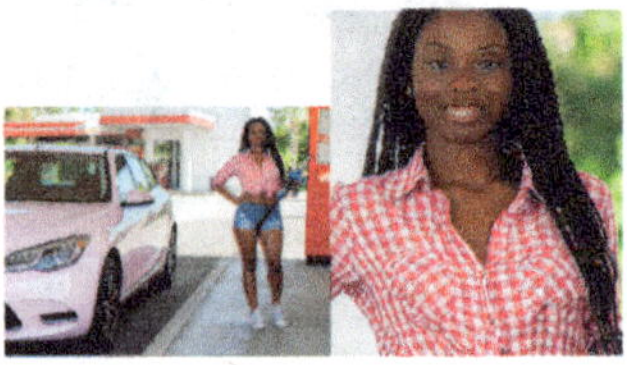

19 button down shirt, denim shorts

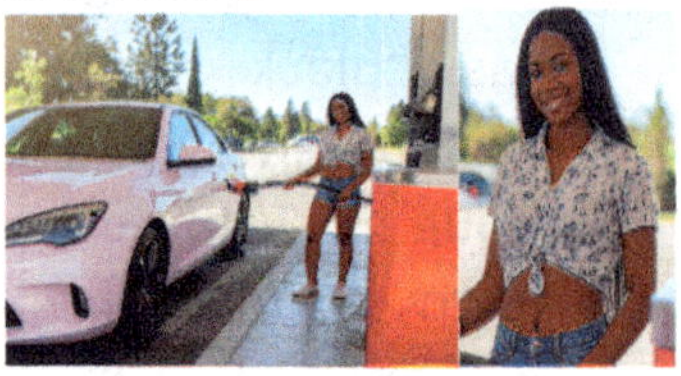

having fun

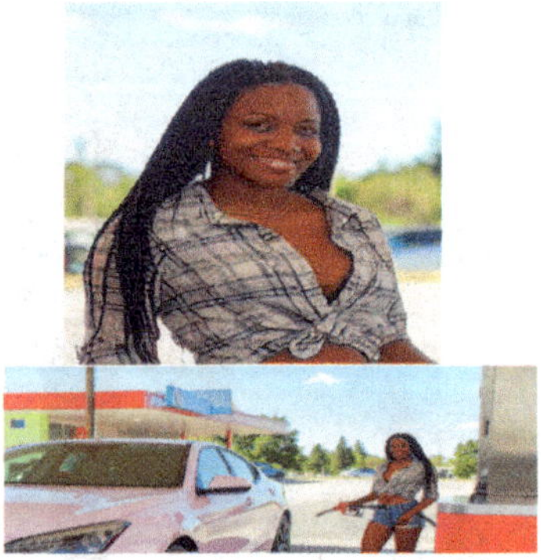

fun in the rain

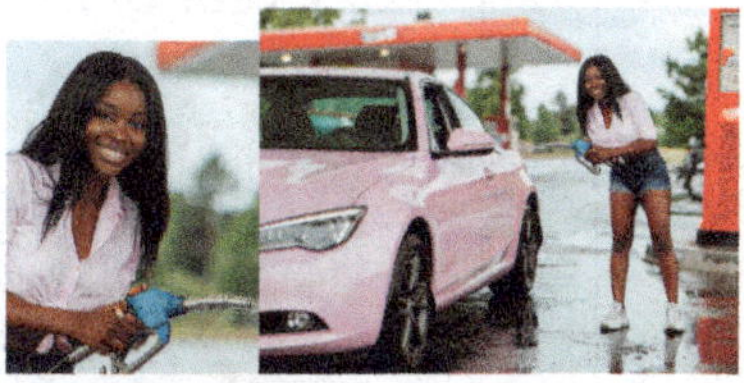

60 year old

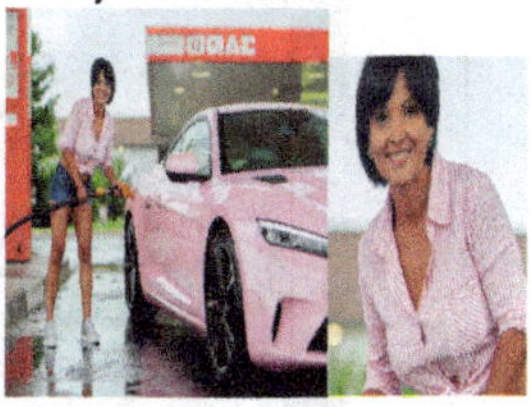

When I asked for "puddles" the models were not made wet. Rain did not come.

50 year old

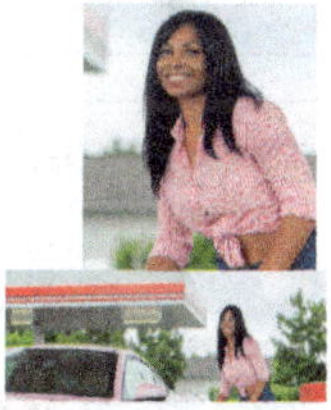

with on leg kicking puddle

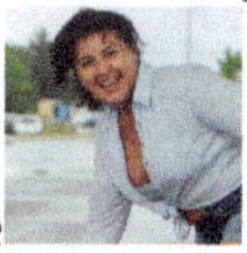

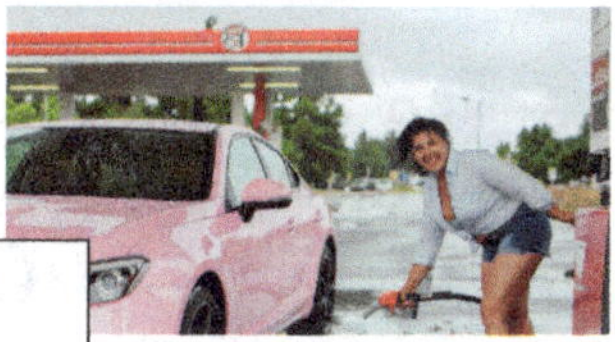

40 year old

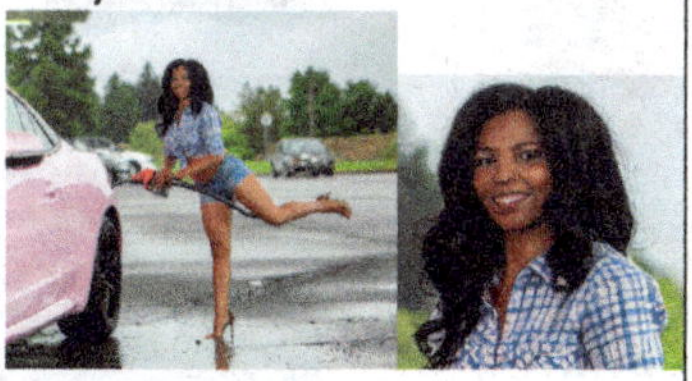

35 year old

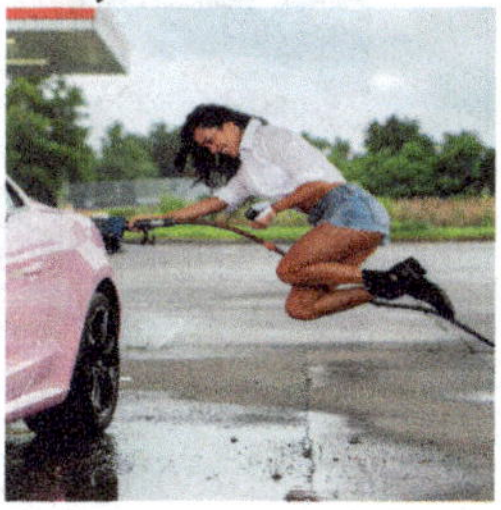

35 year old

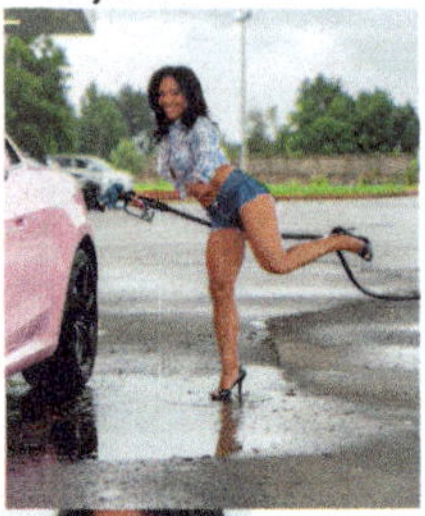

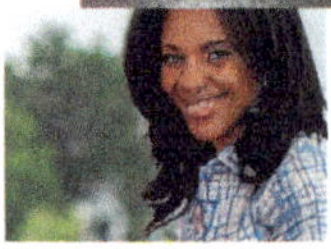

27 year old

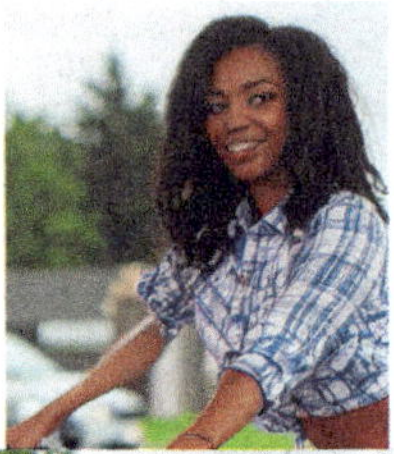

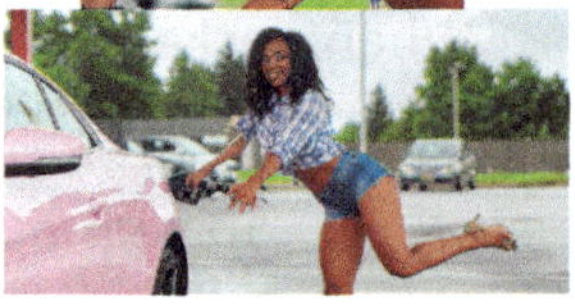

into a rainbow colored convertible

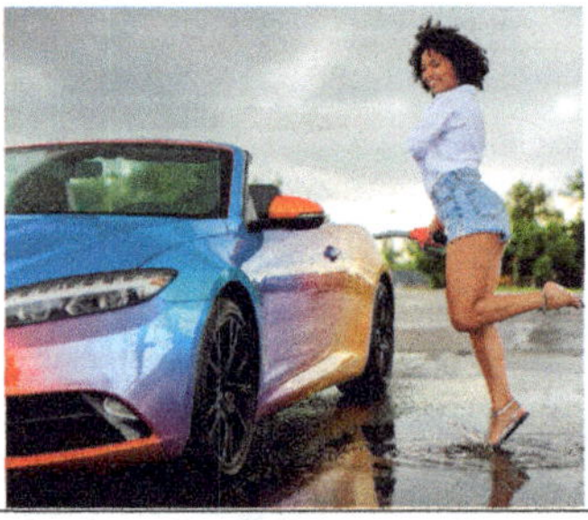

27 year old

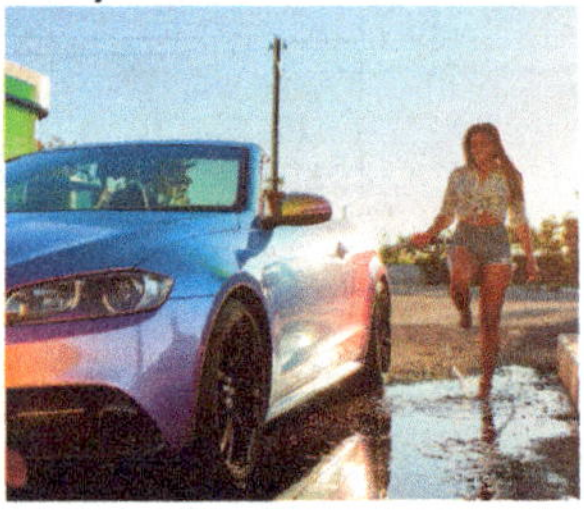

in button down shirt, denim shorts, having fun in the california sunshine with one leg kicking puddle

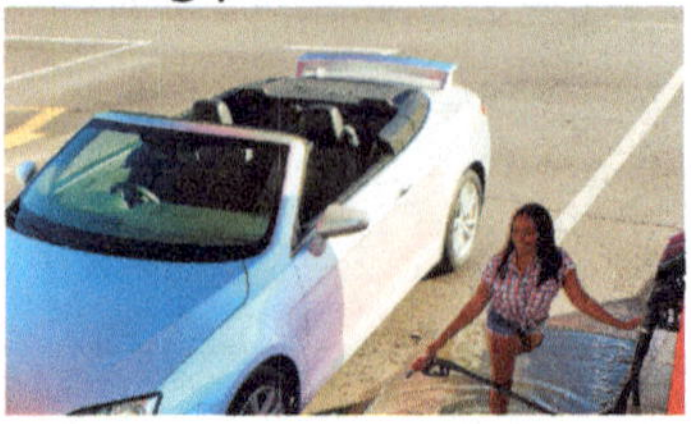

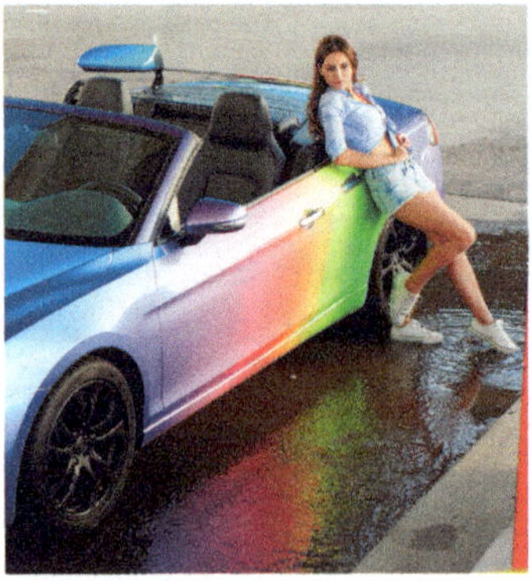

27 year old brown

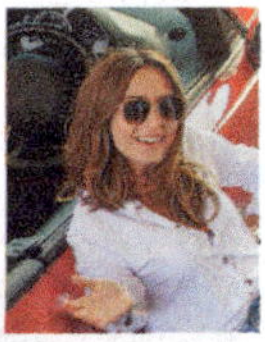

headed

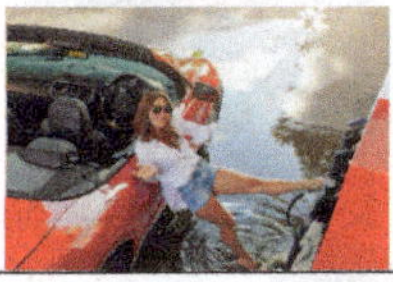

26 year old

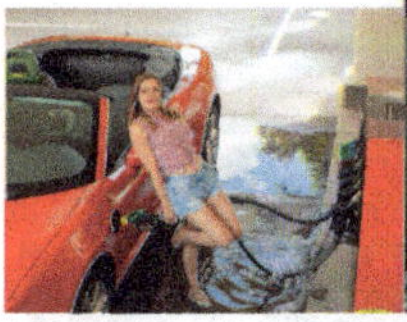

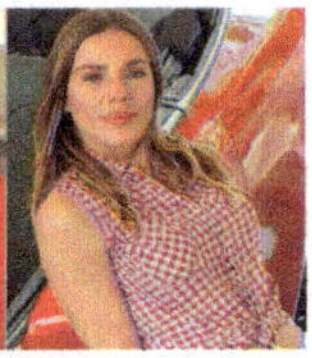

25 year old brown headed

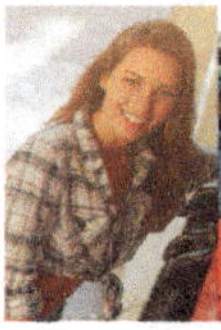

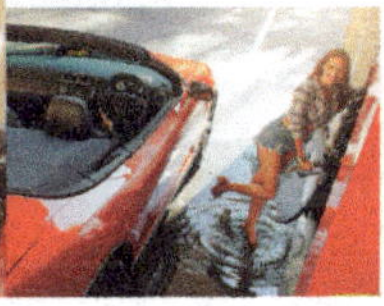

24 year old kicking puddle

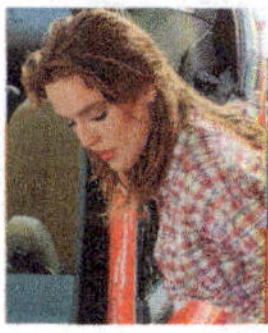

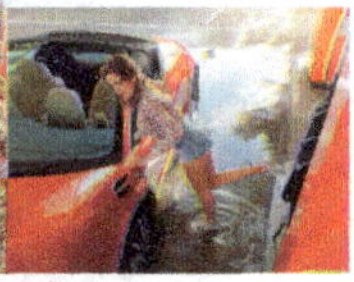

23 year old

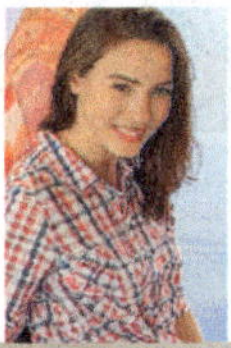

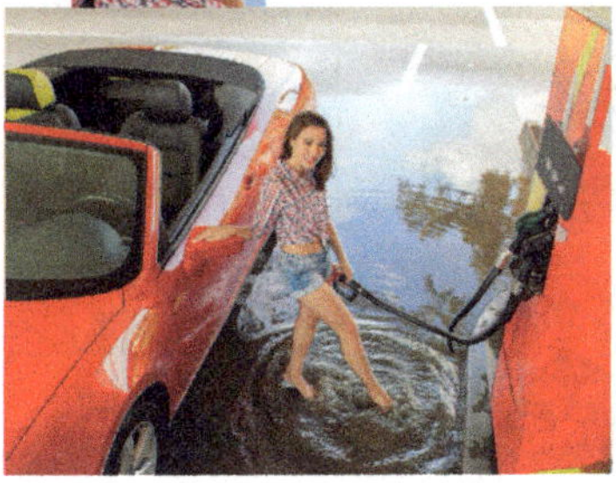

22 year old

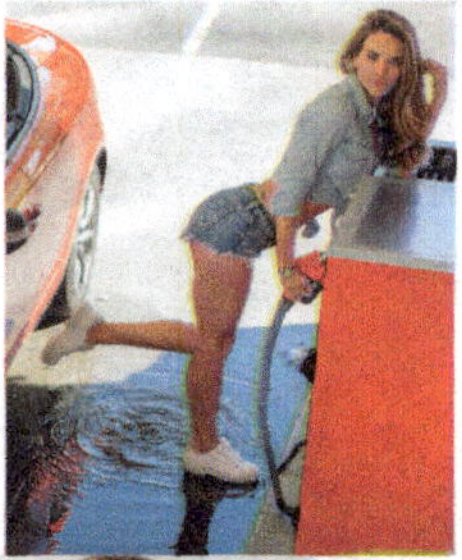

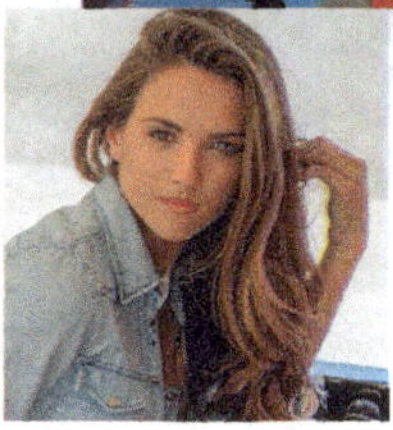

21 year old

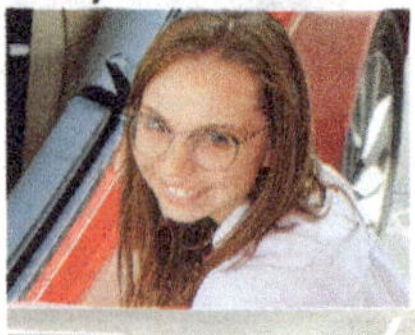

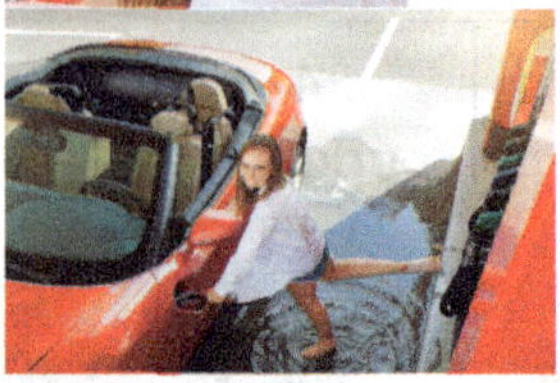

18 year old

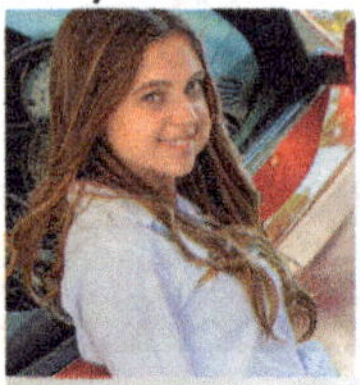

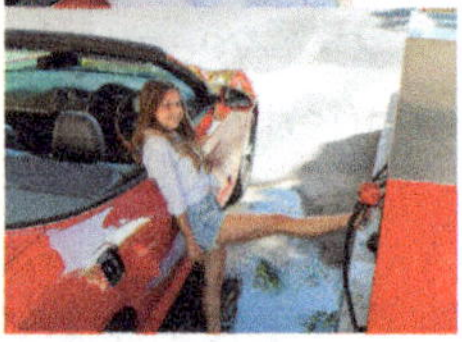

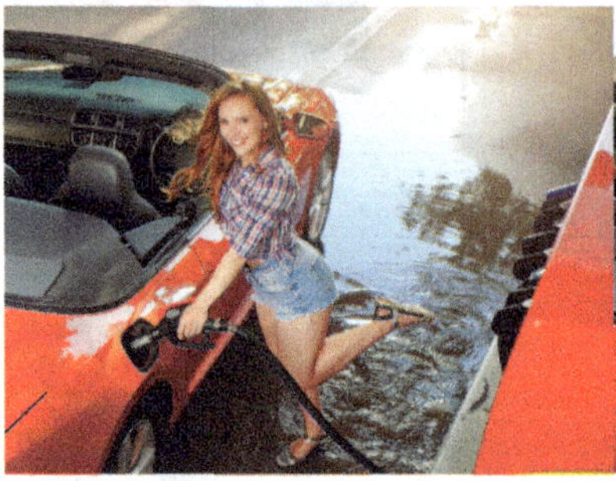

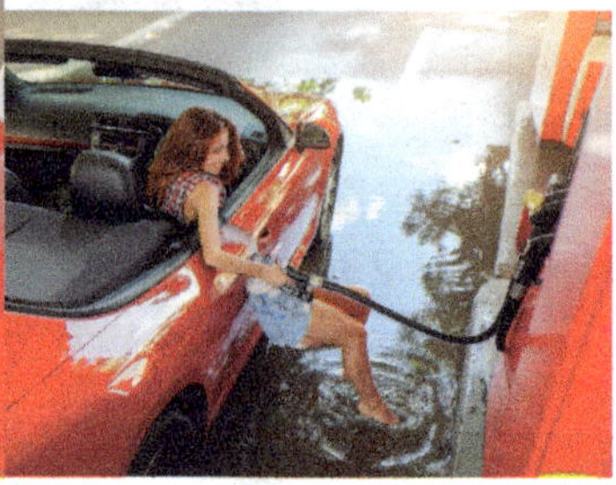

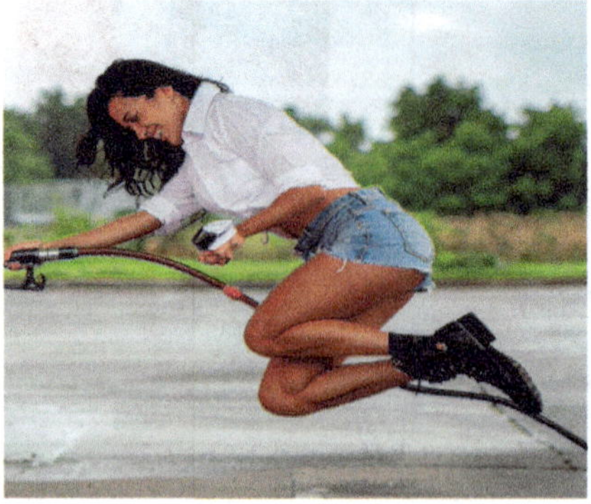

Pretty amazing how many hoses went nowhere or went through bodies or were split

18 black colored convertible

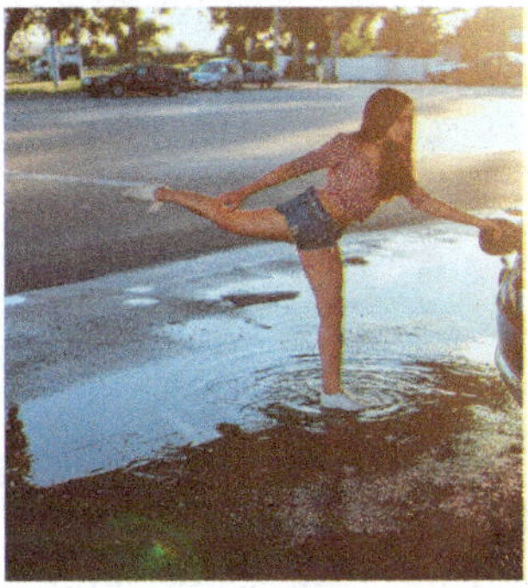

19 year old brown headed

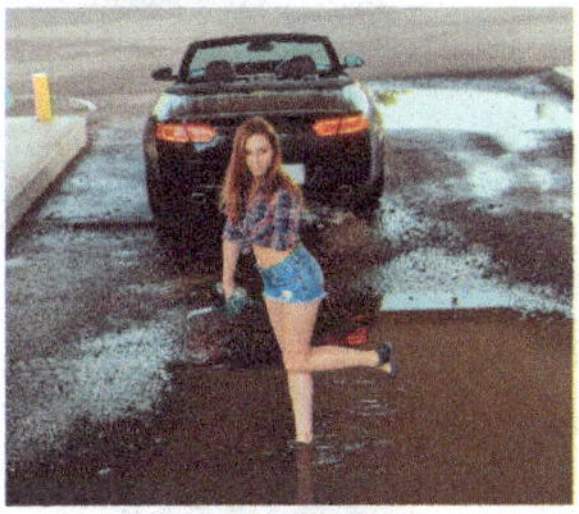

20 year old

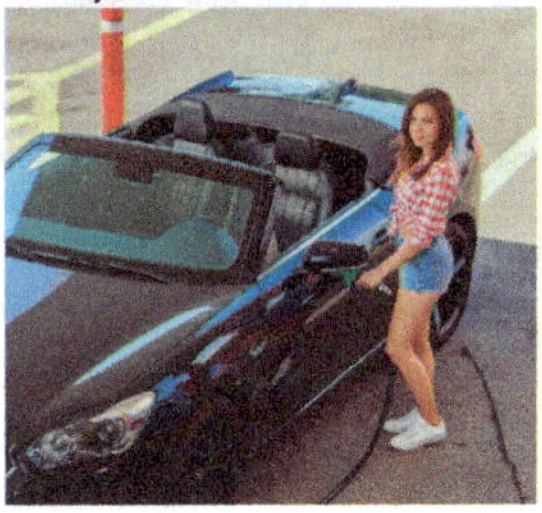

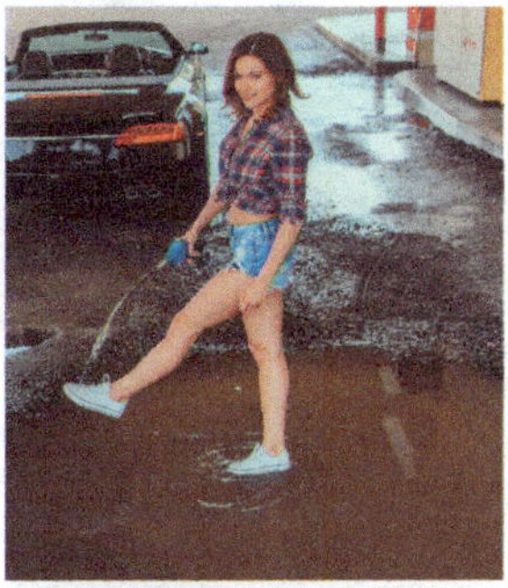

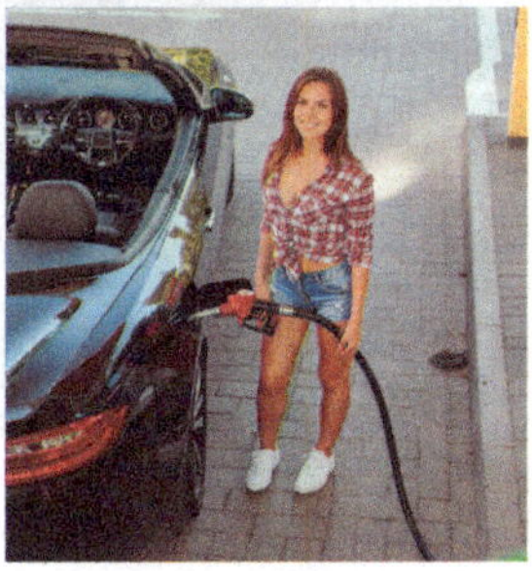

21 year old

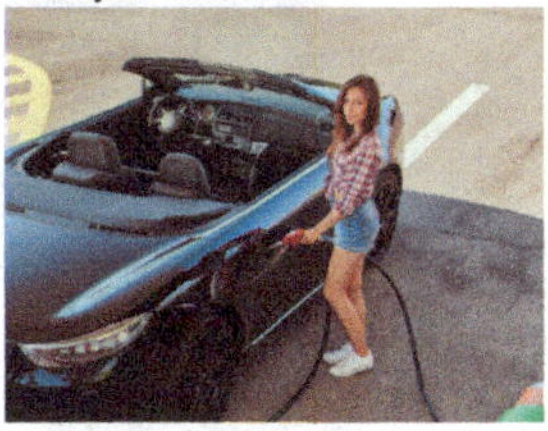

21 year old

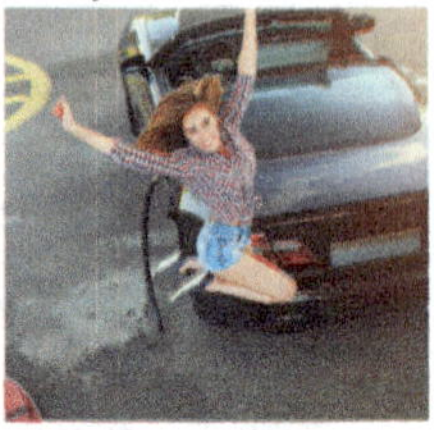

21 year old

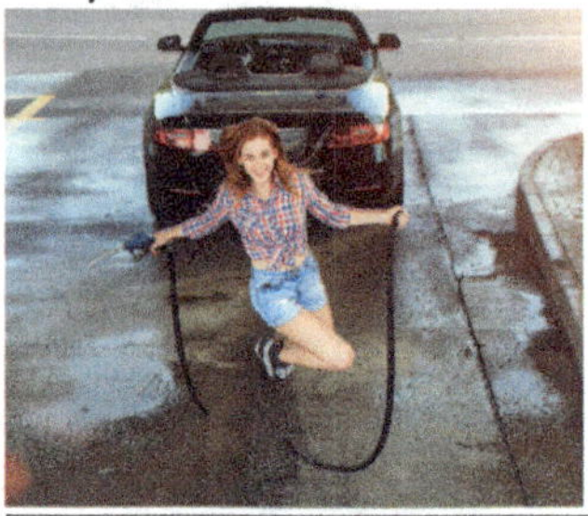

22 year old

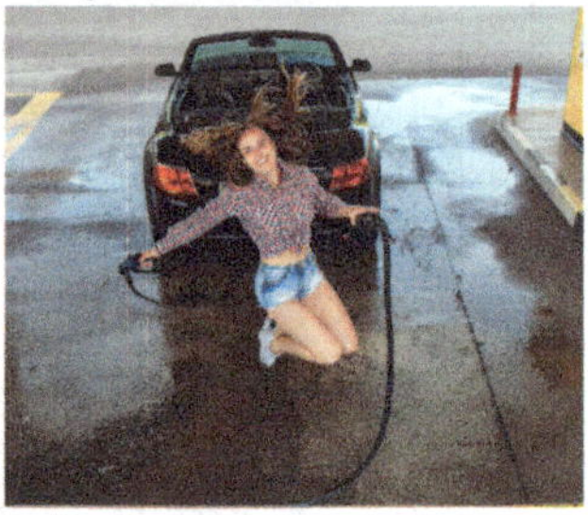

22 year old brown in hand

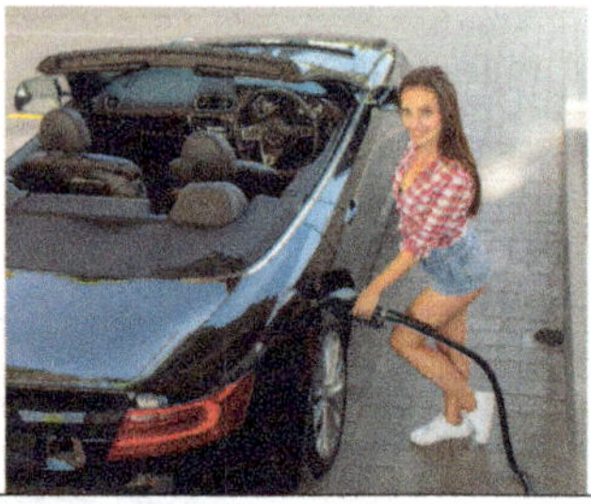

23

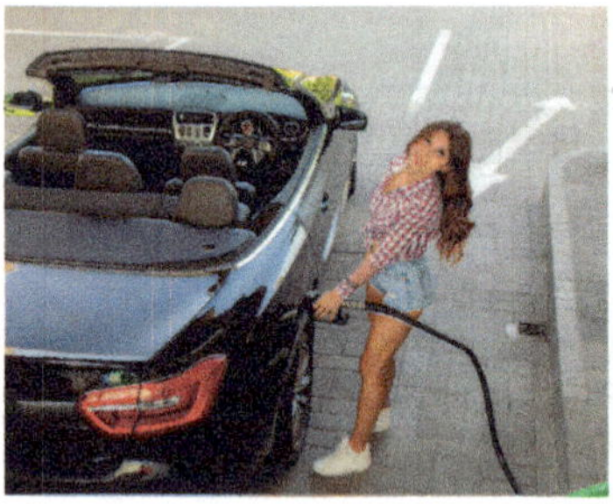

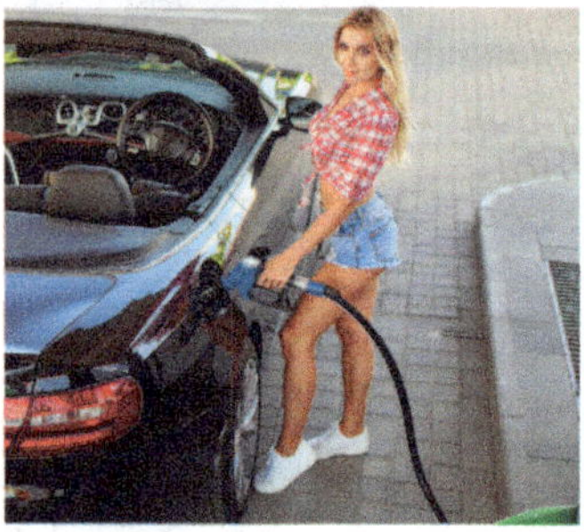

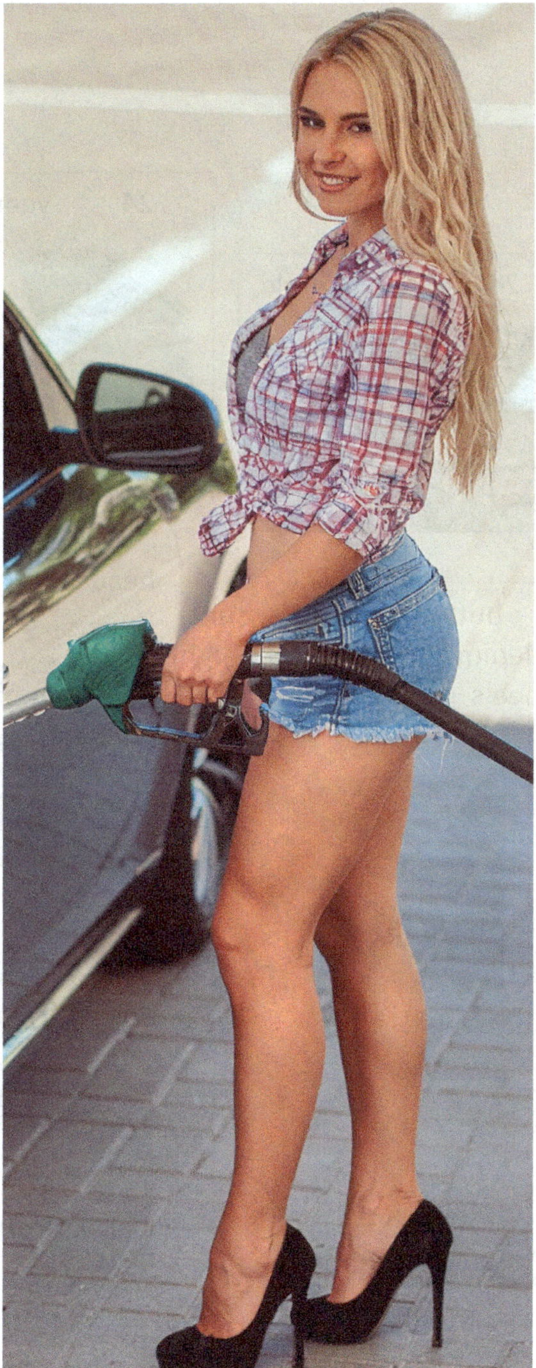

24 year old

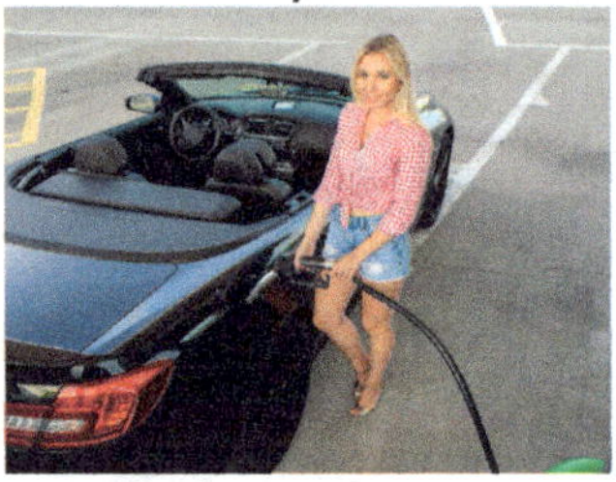

24 year old

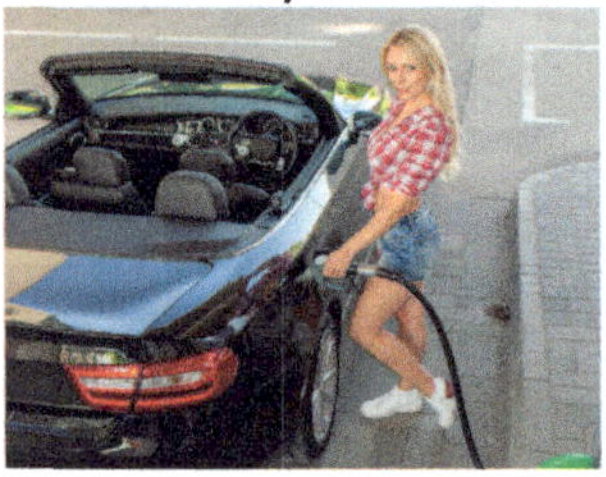

24-year-old red headed

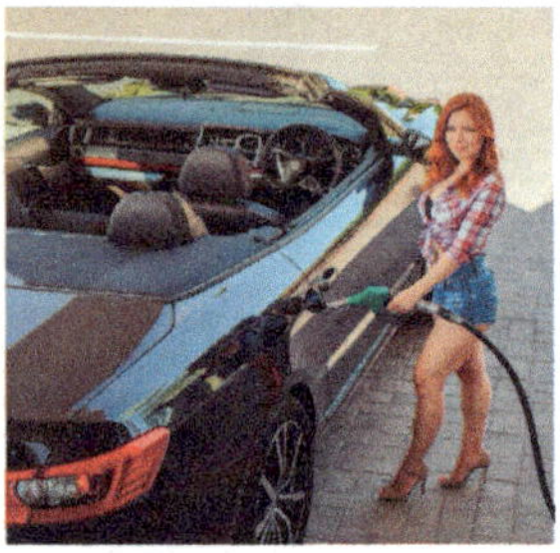

button down shirt, denim shorts and high heels posing

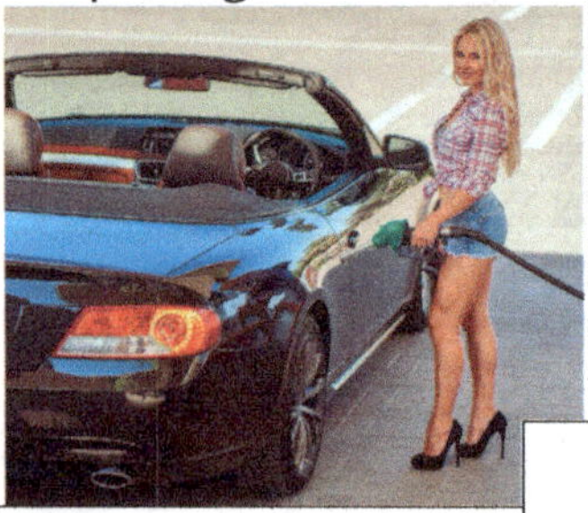

25 year old red headed posing with smile

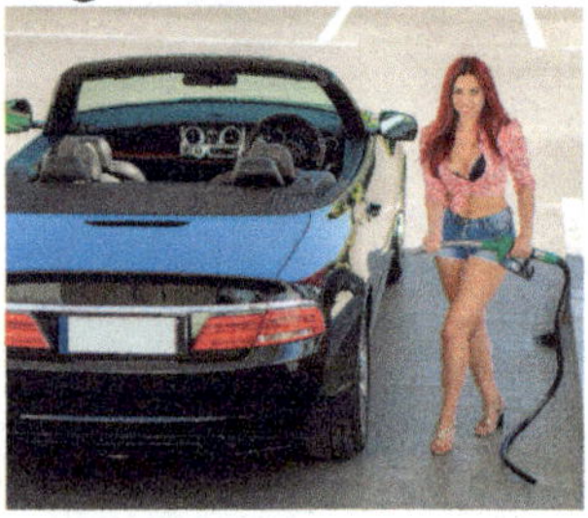

25 year old

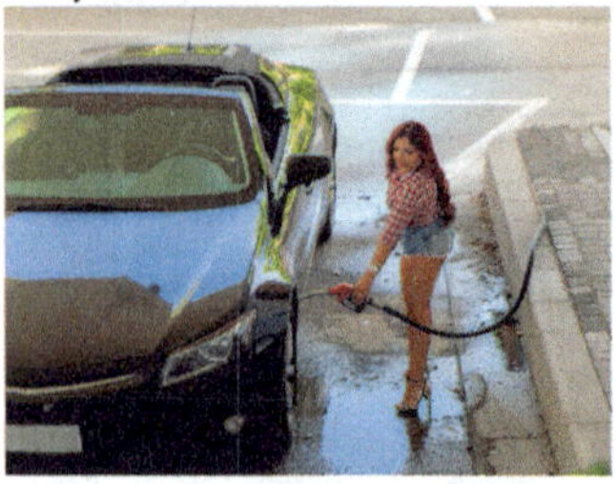

full body shot from directly above, of sexy 24 year old blond headed woman at a gas station, with hose in hand, pumping gas into a black colored convertible sports car in button down shirt, denim shorts and high heels posing

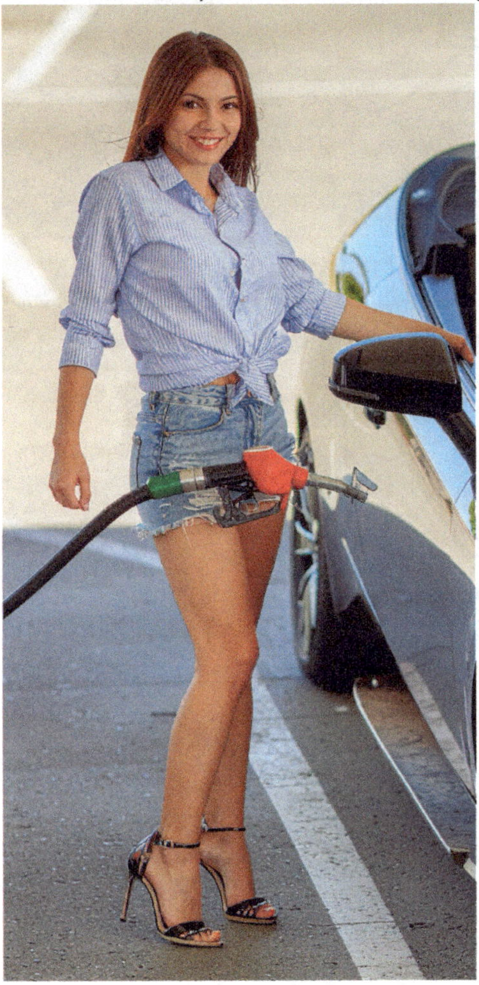

26 year old posing with smile

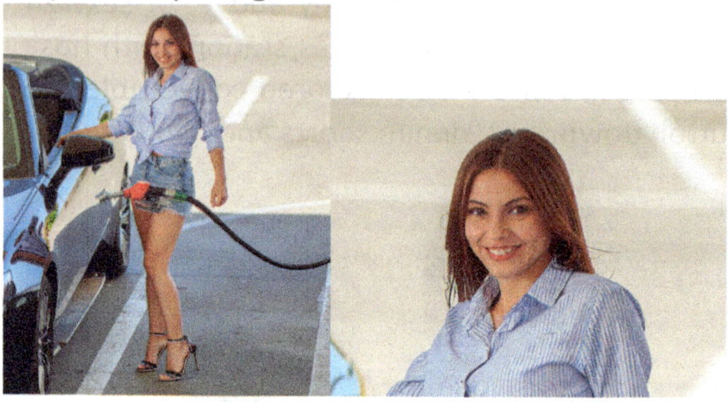

26 year old red headed woman

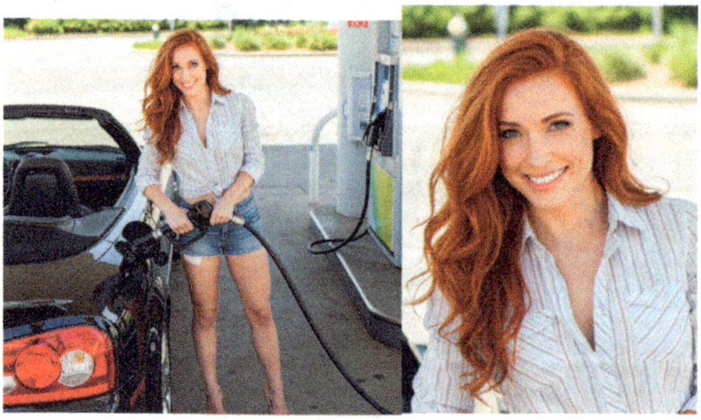

19 year old

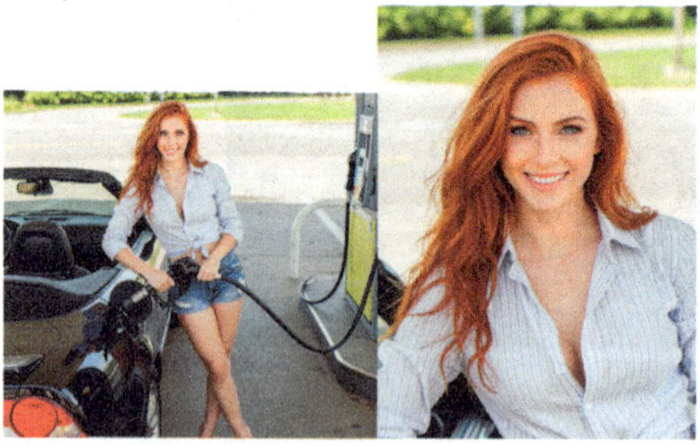

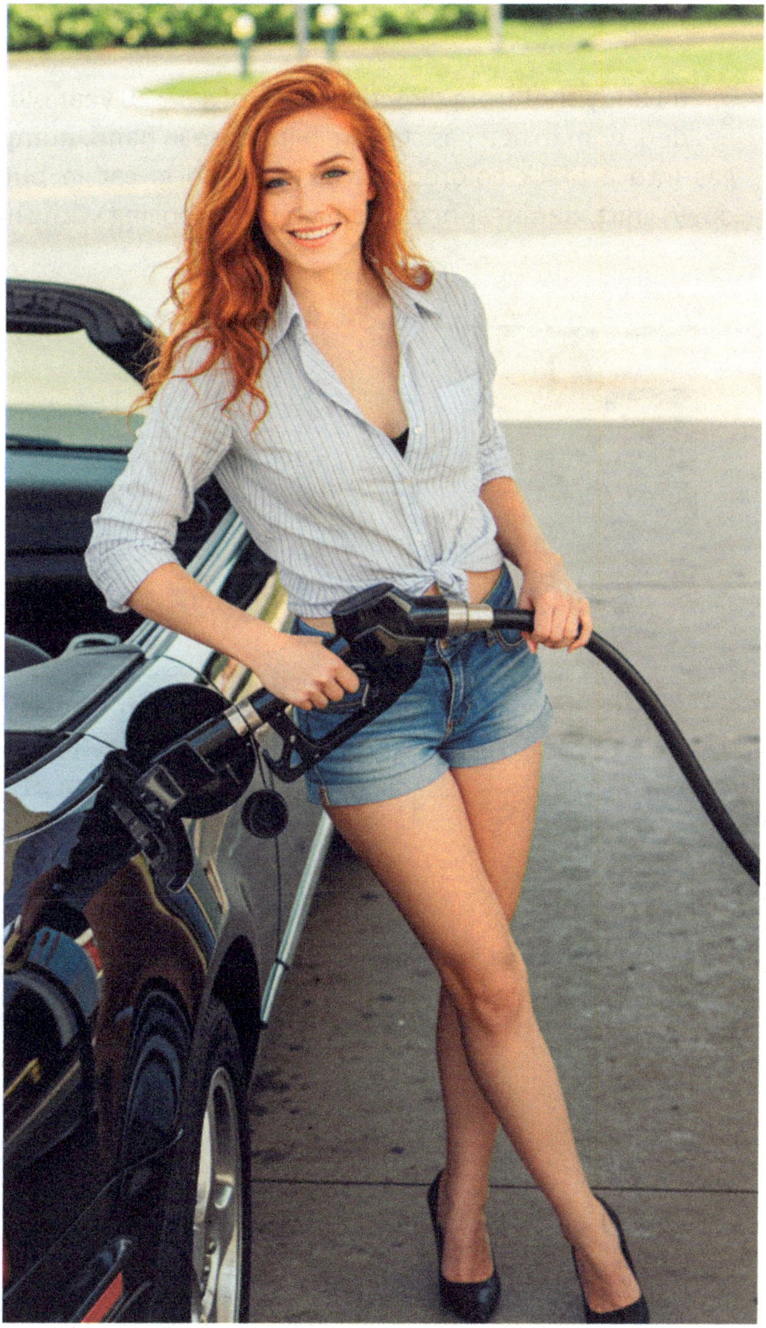

full body shot from directly above, of sexy 20 year old red headed woman at a gas station, with hose in hand, pumping gas into a black colored convertible sports car in button down shirt, denim shorts and high heels posing with smile

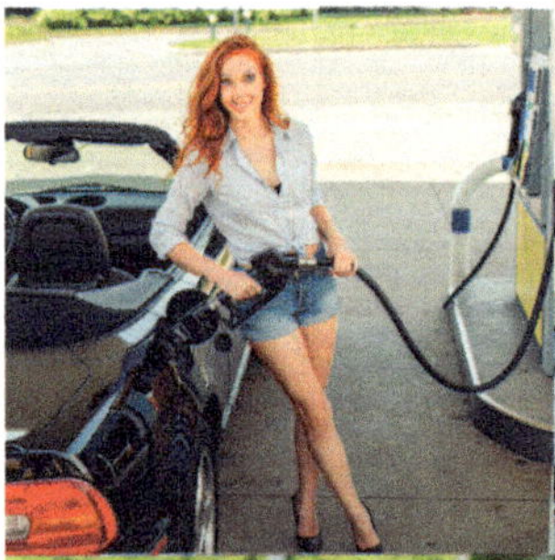

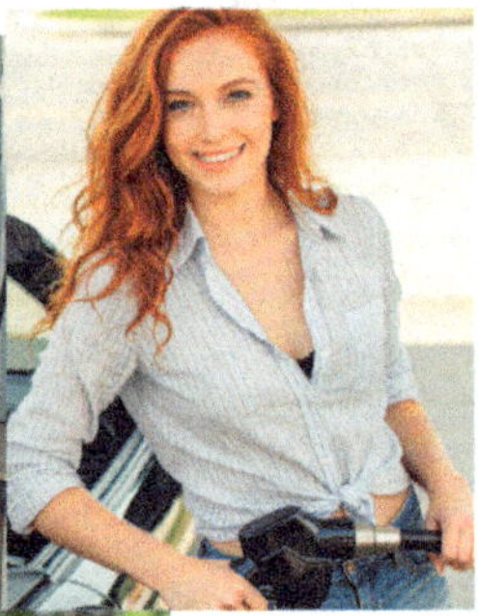

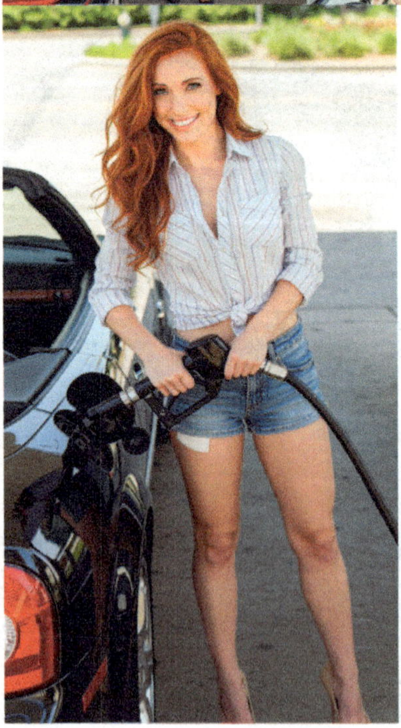

I haven't been able to see how well differences and age of model prompts were handled by the program. I am amazed how many times the same exact prompts came up with similar looks and faces and then, all of a sudden, changed completely.

sexy 20 year old

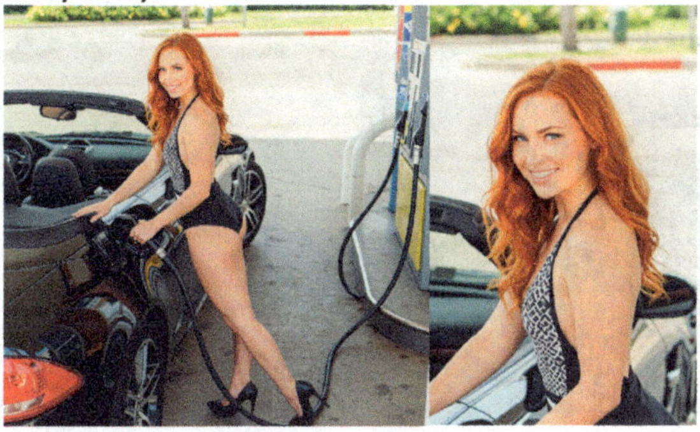

18 year old

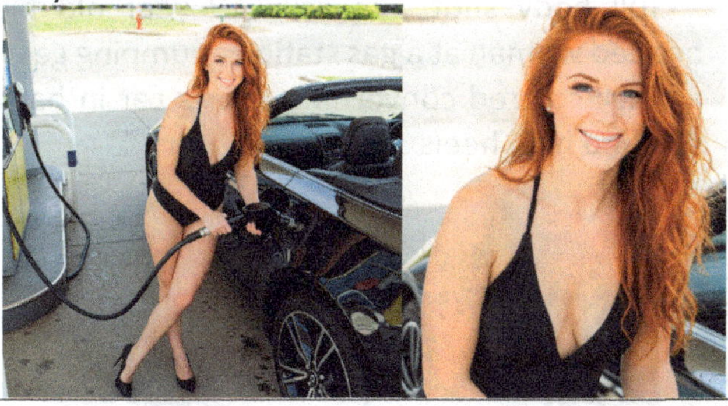

18 year old blond

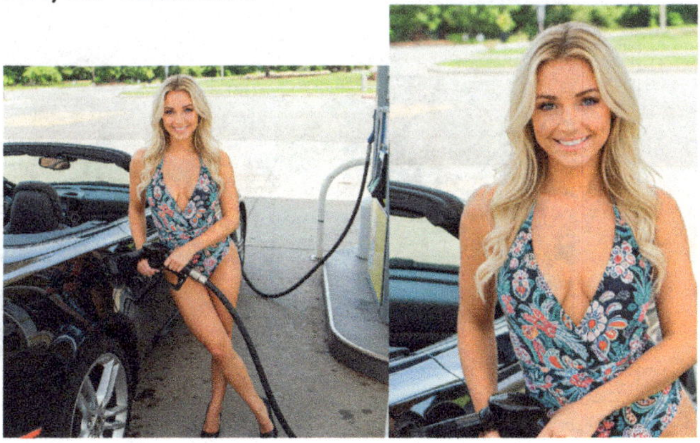

18 view from behind

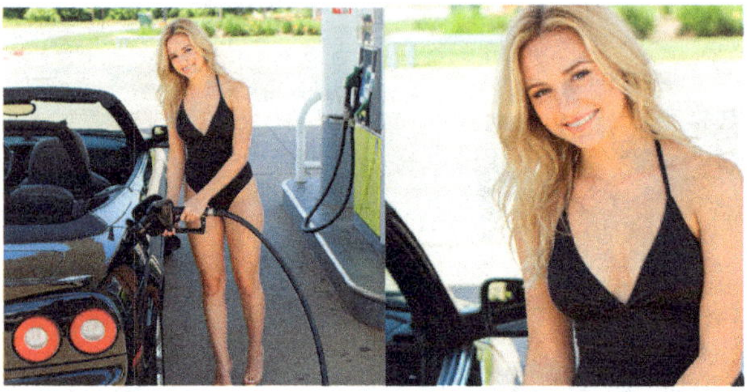

full body shot , of sexy 19 year old blond headed woman at a gas station, pumping gas into a black colored convertible sports car in bathing suit and high heels

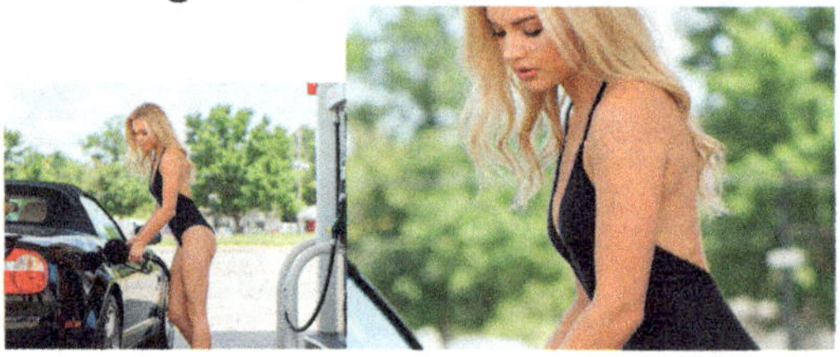

20 year old

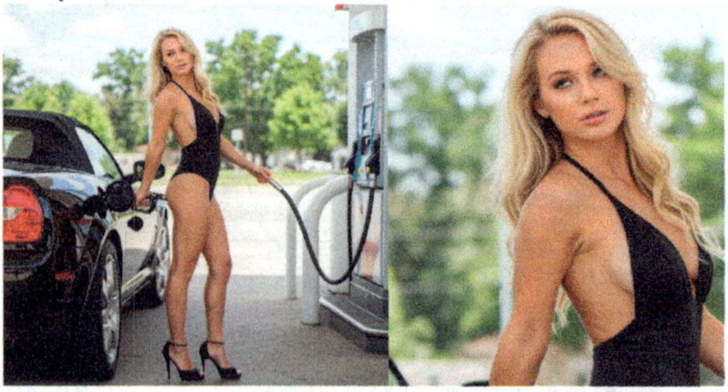

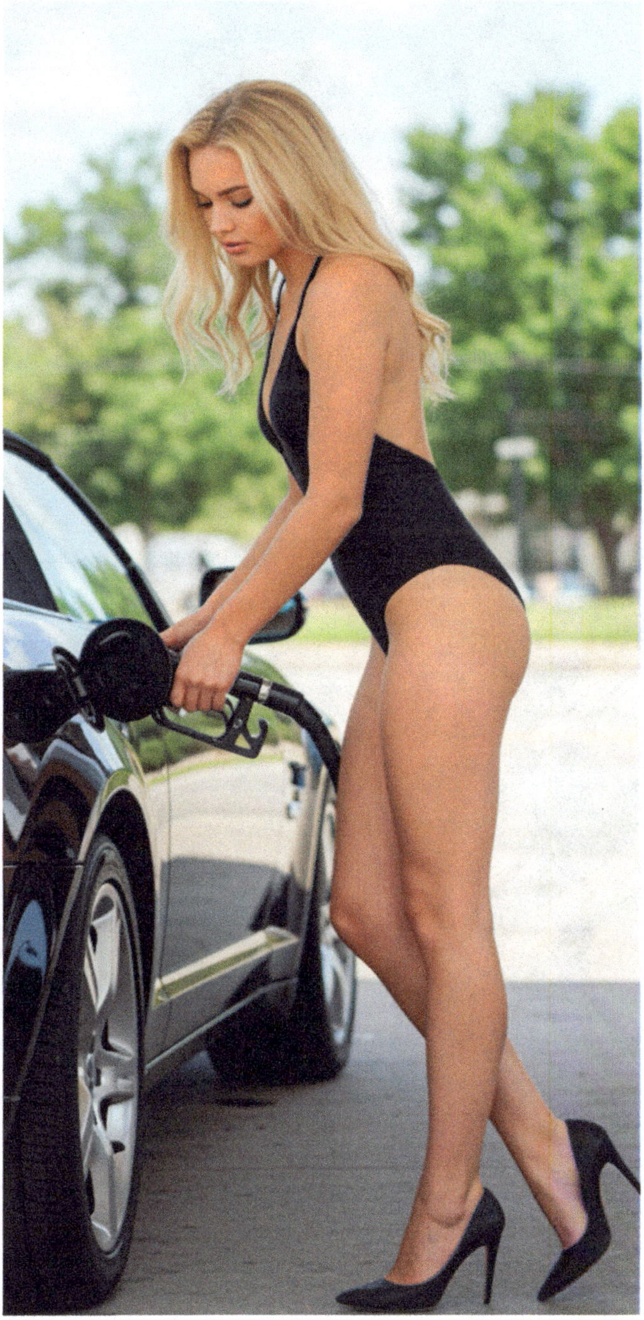

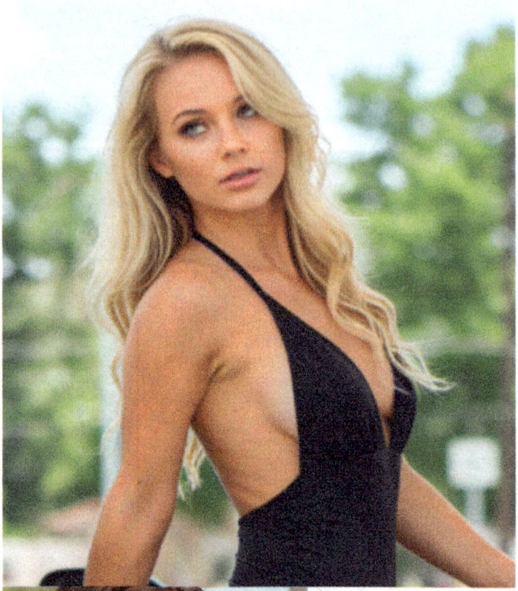

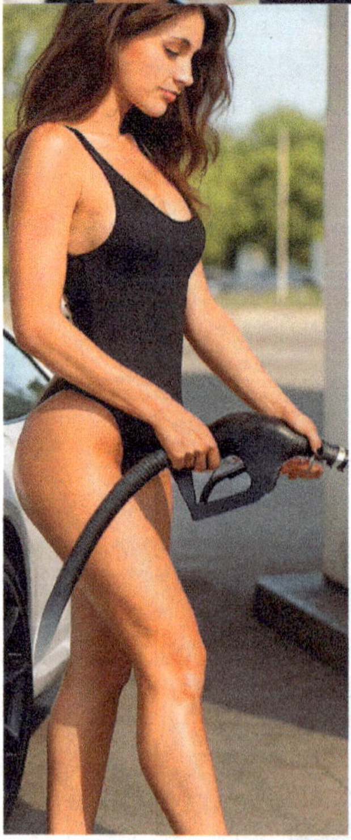

woman pumping gas at gas station

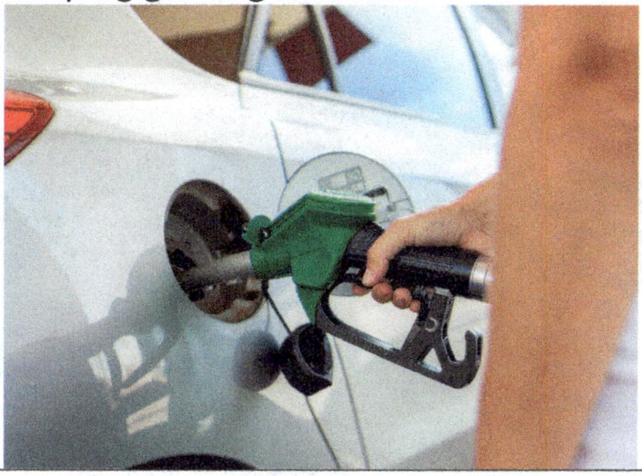

full body shot of woman pumping gas at gas station

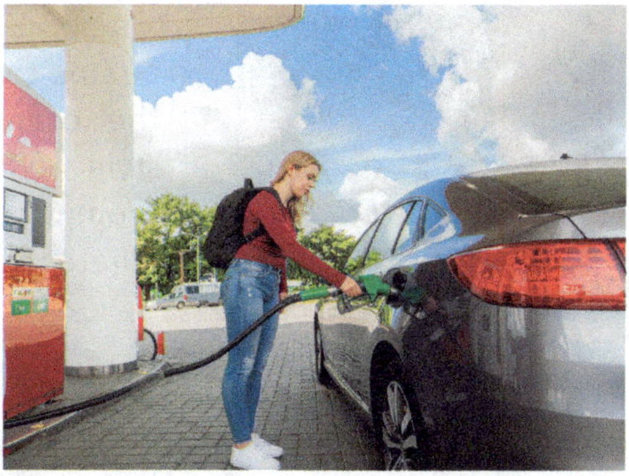

full body shot of woman pumping gas at gas station high heels

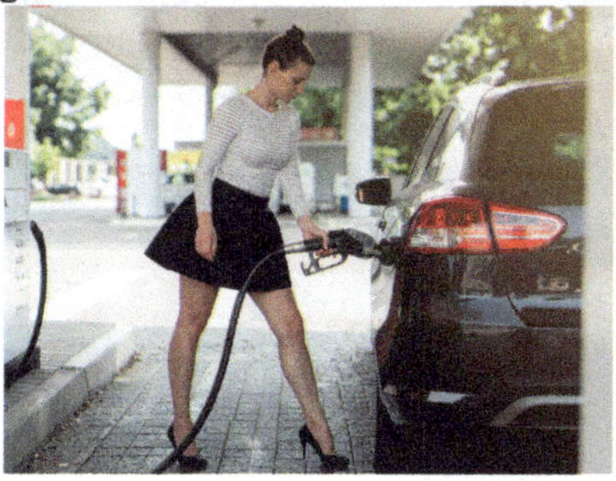

full body shot of woman pumping gas at gas station in high heels

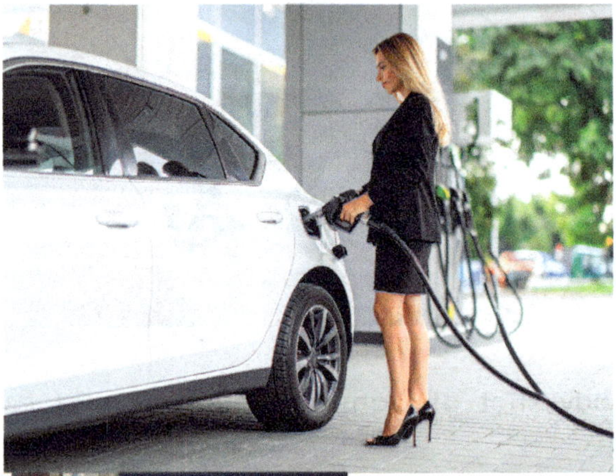

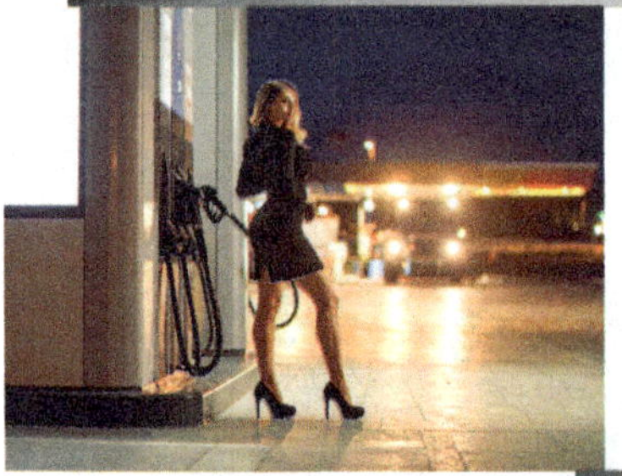

One can actually see the change in image style when similar prompts are put into different AI programs.

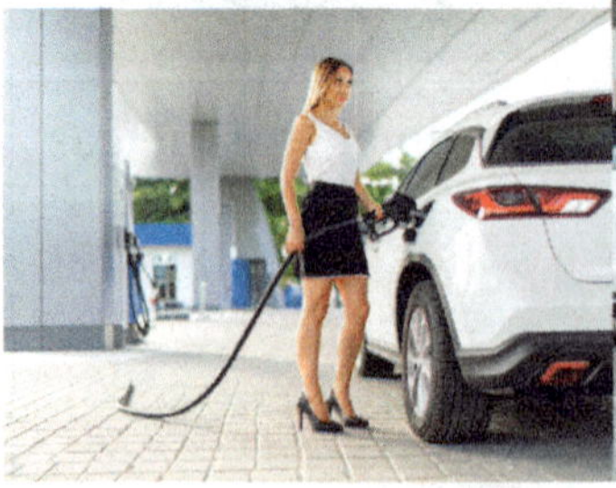

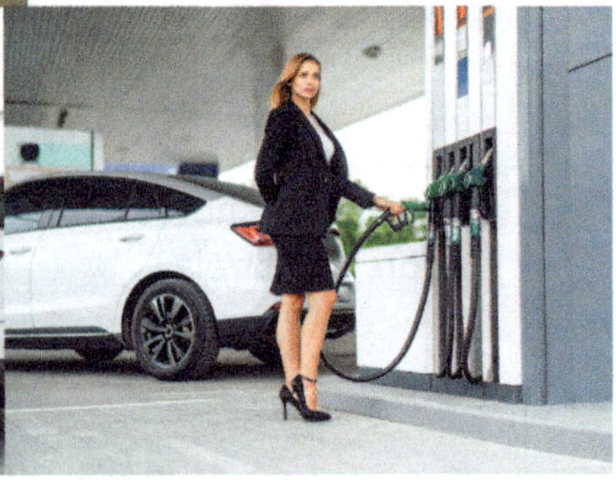

full body shot of woman in high heels and bathing suit pumping gas at gas station

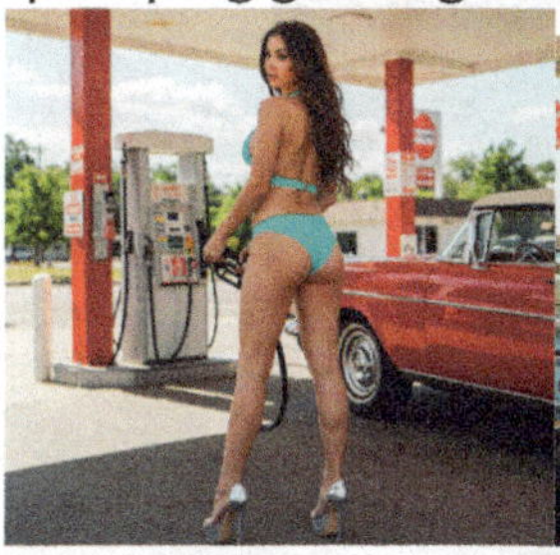

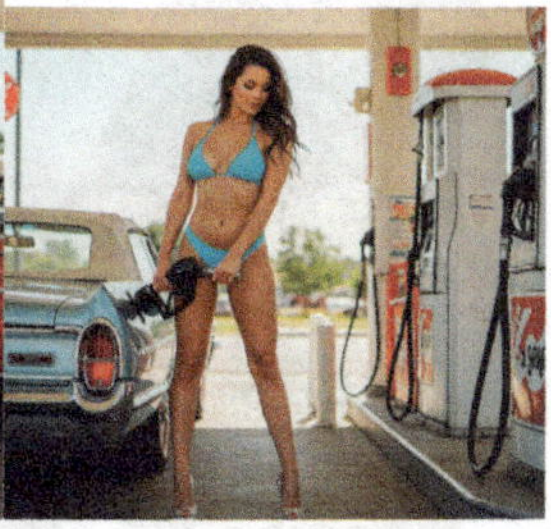

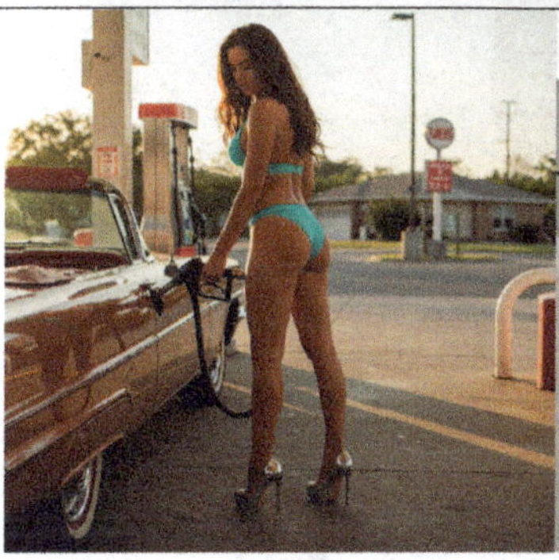

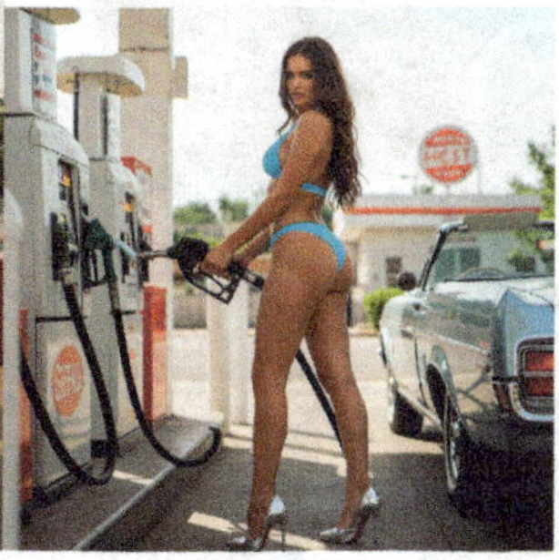

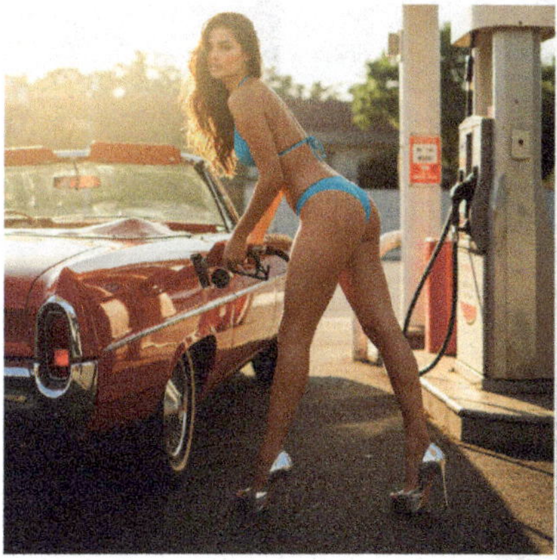

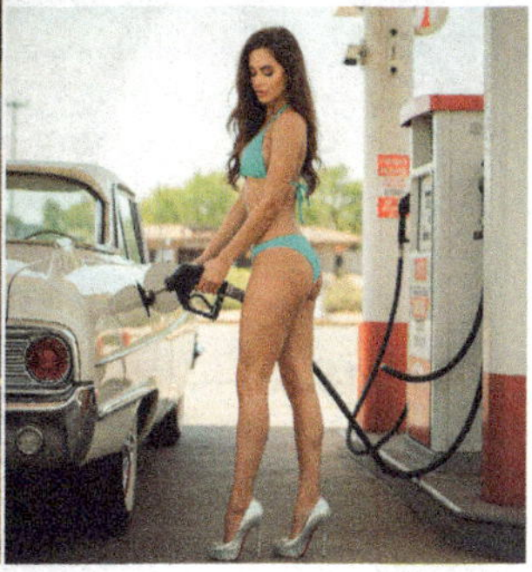

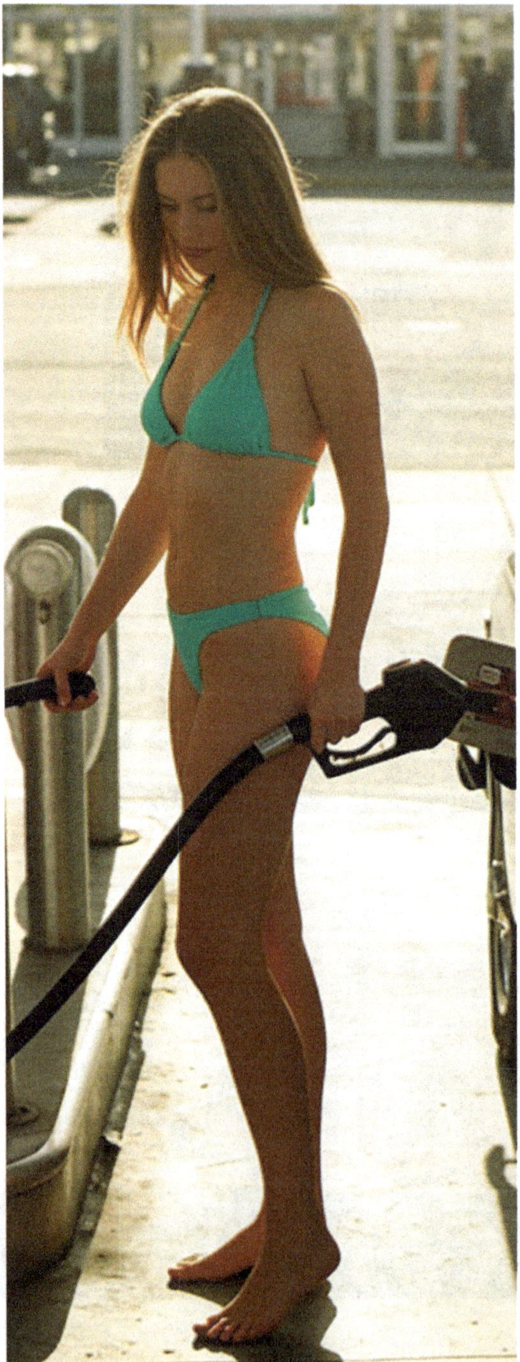

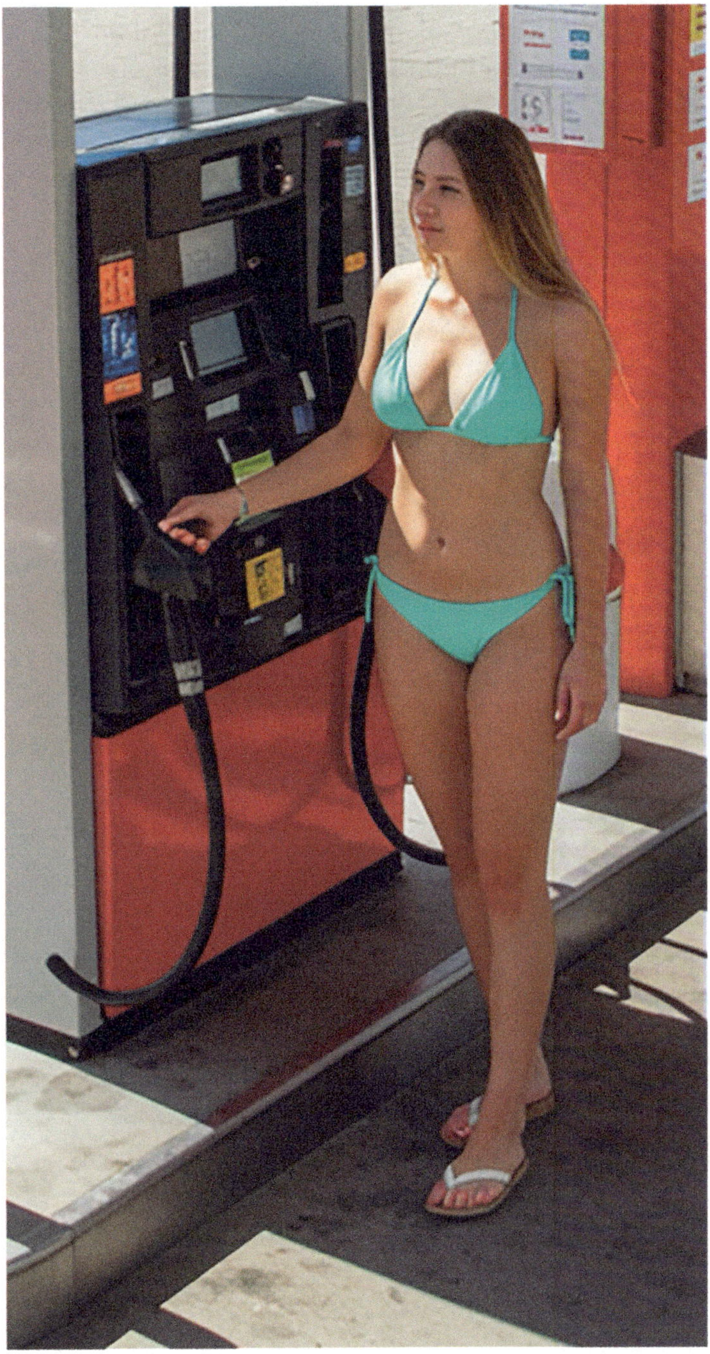

full body shot of woman with black hair in high heels and bathing suit pumping gas at gas station

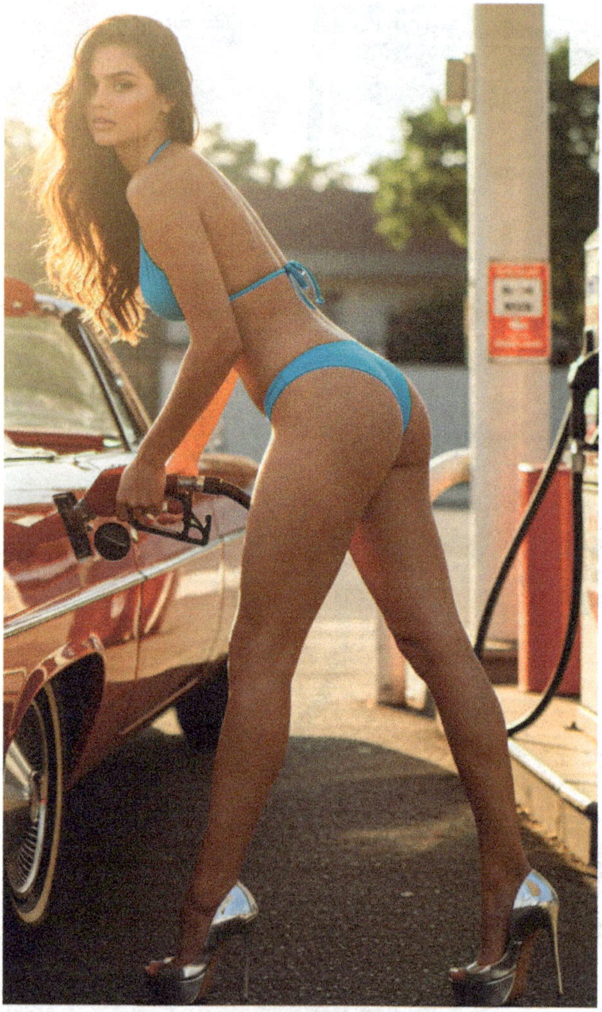

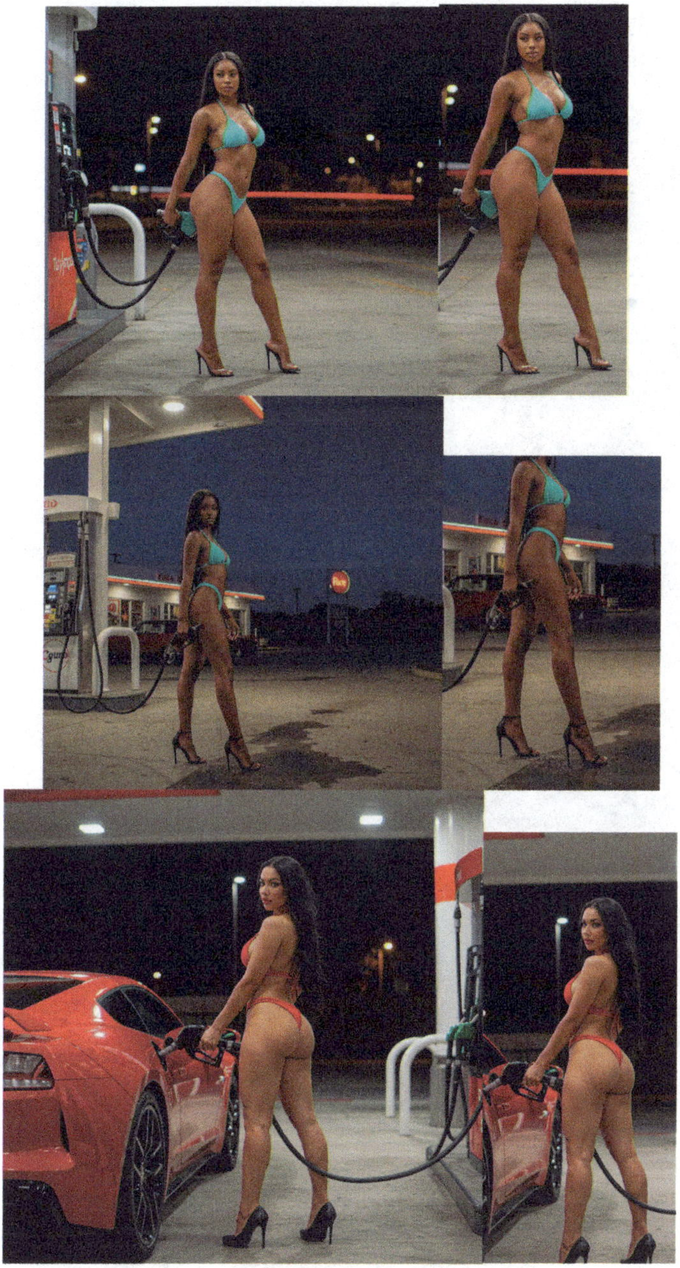

full body shot of woman with black hair in high heels and bathing suit pumping gas into sports car at gas station

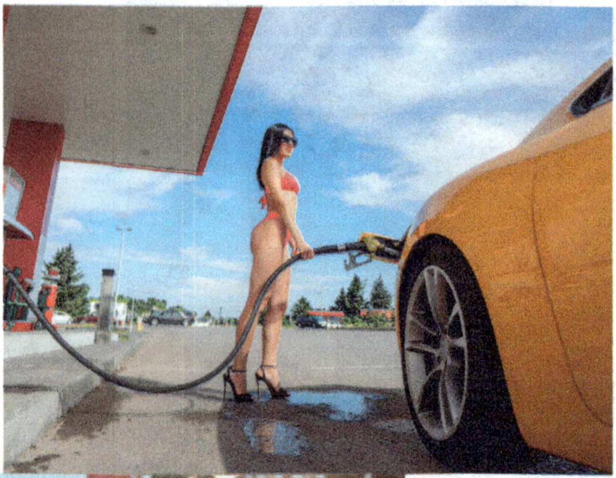

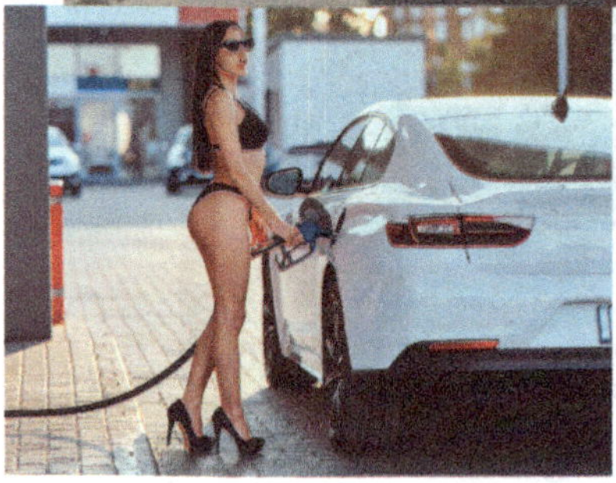

I saw instances of uncalled for changes in light, darkness, and, as you will see, sometimes, bad attempts at artsy fartsy lighting..

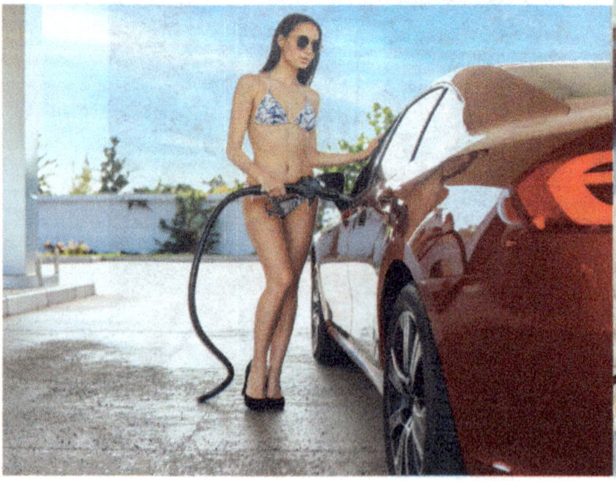

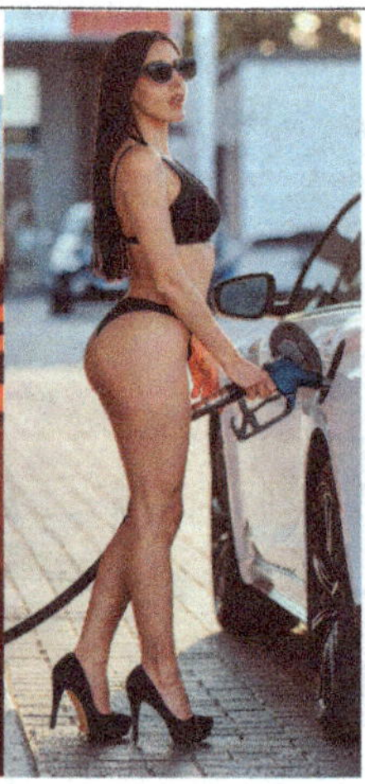

full body shot of woman with black hair in high heels and bathing suit pumping gas into sports car at gas station

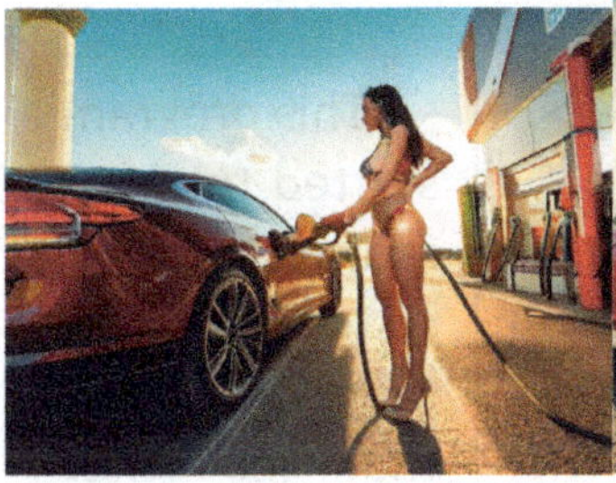

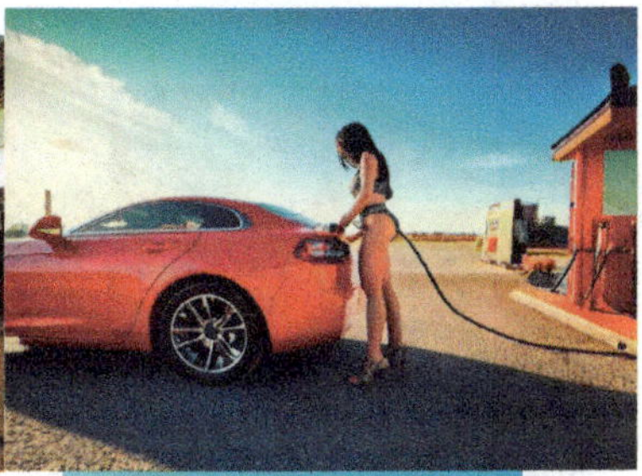

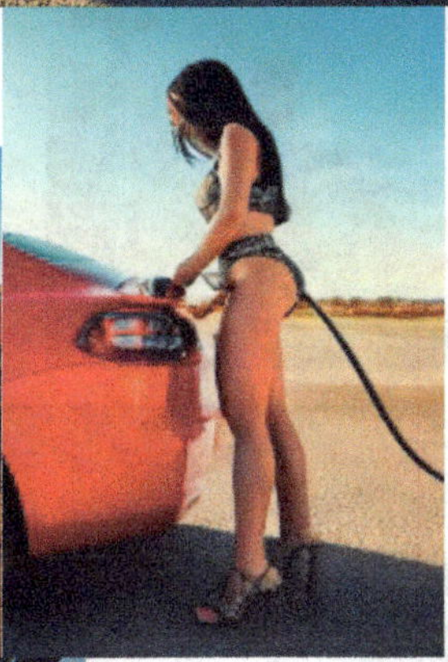

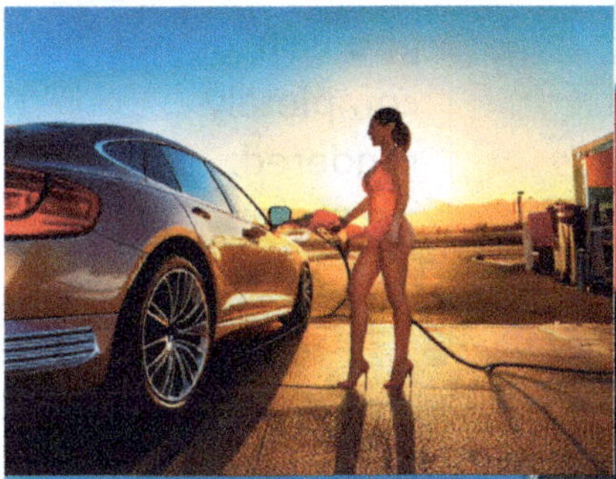

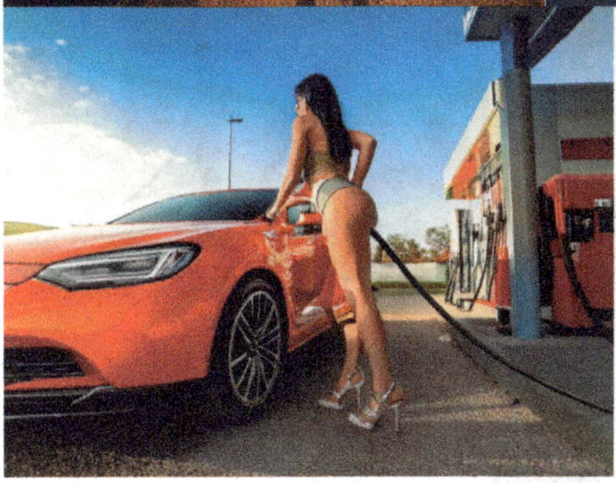

Why the backlight of glary sunshine from this program? Why a dessert landscape?

full body shot of woman with red hair in high heels and bathing suit pumping gas into sports car at gas station

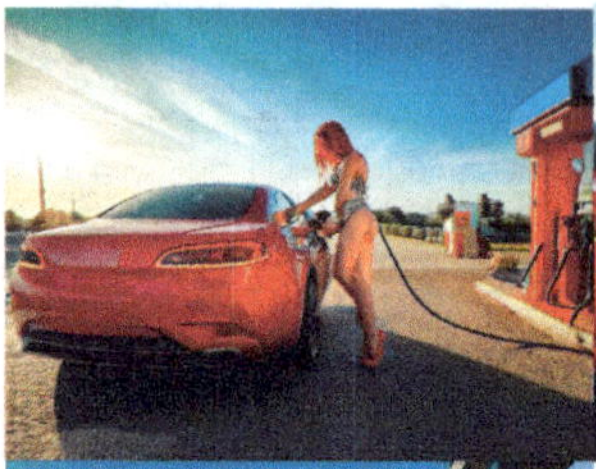

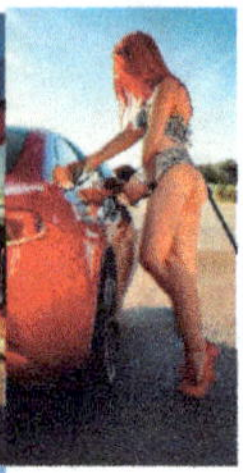

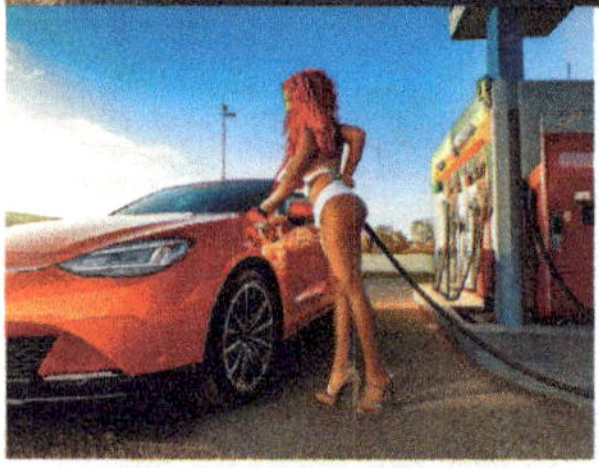

This program started to show radical distortions more often than the other programs and faces that were ghoulish or not completely rendered.

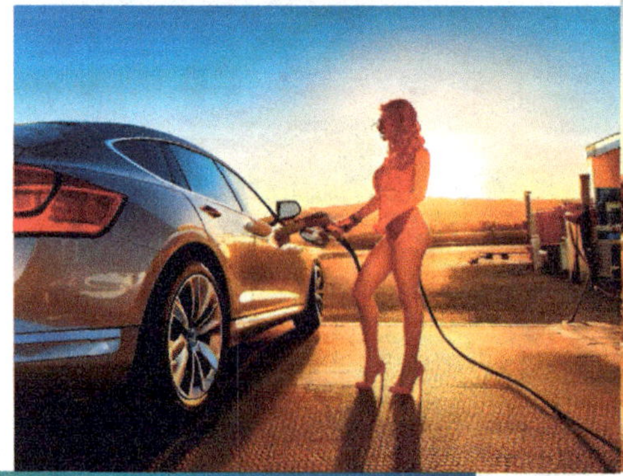

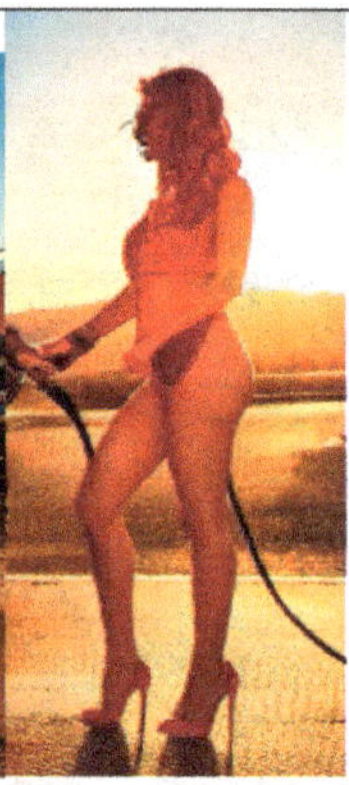

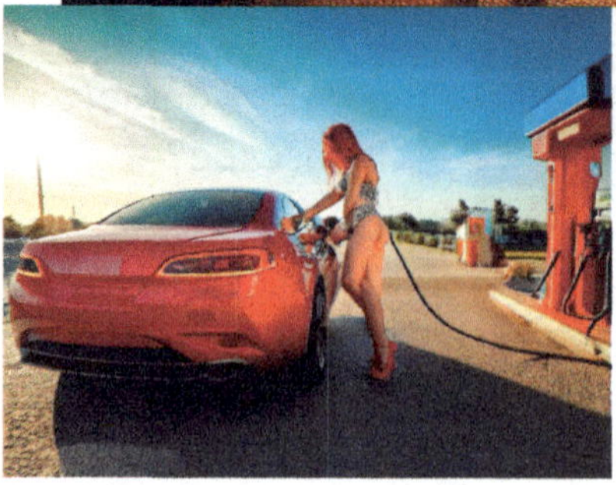

full body shot of woman with red hair in high heels and bathing suit pumping gas into sports car at gas station

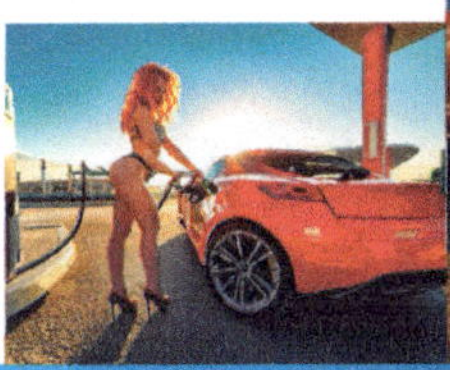

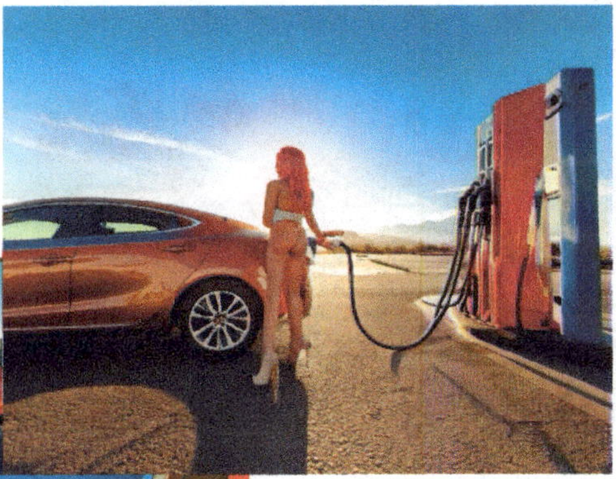

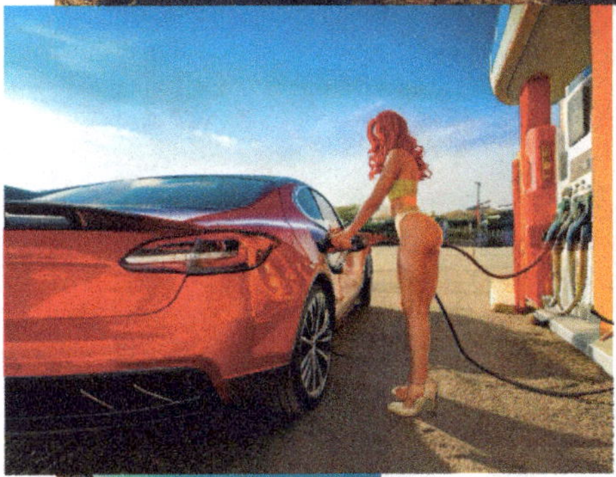

Here we find a complete ignorance of details in the human models and some shiny car objects as more of a focus. I really don't understand the lighting or the setting. I didn't ask for it, as you can see by the prompts.

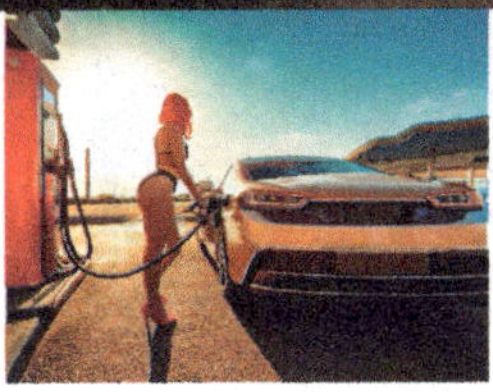

full body shot of woman with blond hair in high heels and bathing suit pumping gas into sports car at gas station

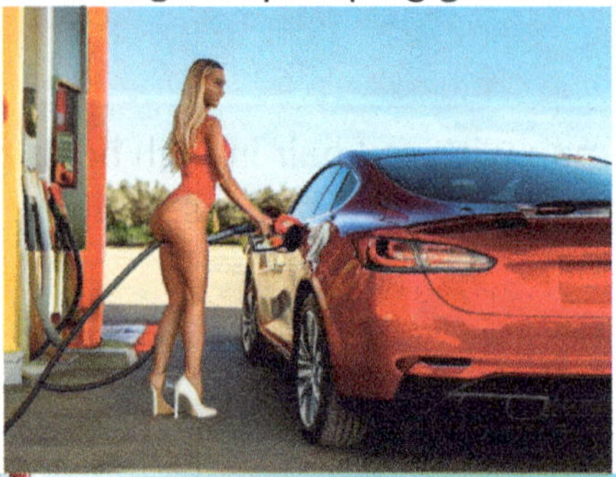

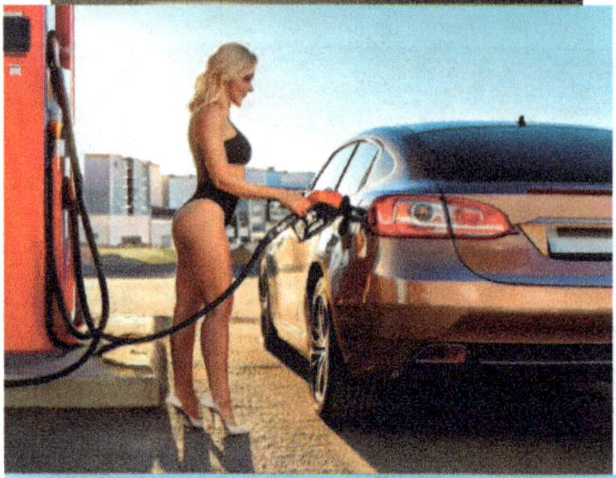

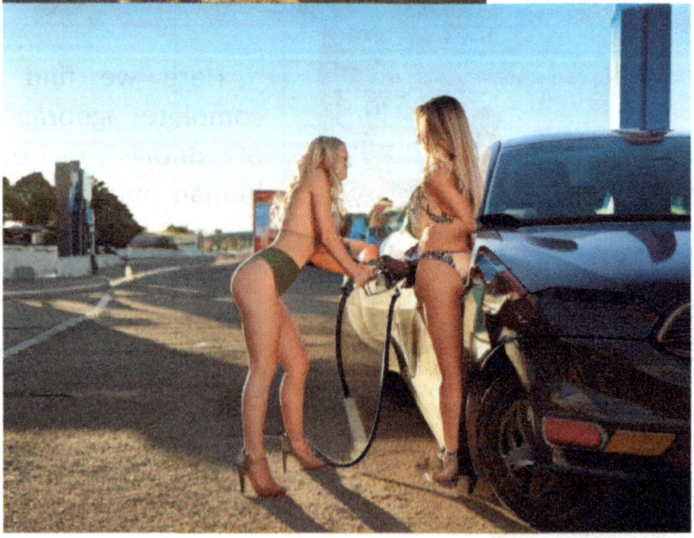

full body shot of woman with blond hair in high heels and bathing suit pumping gas into sports car at gas station

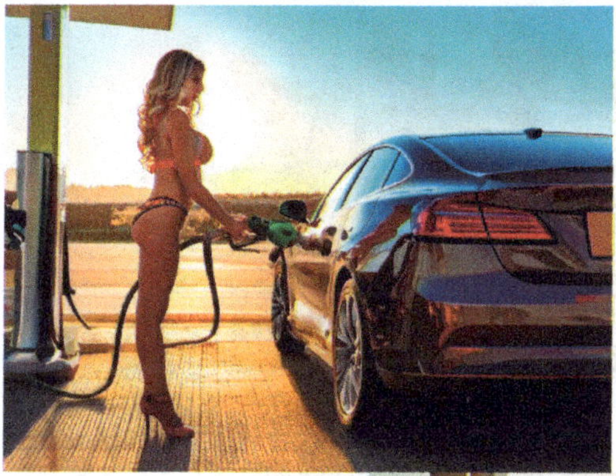

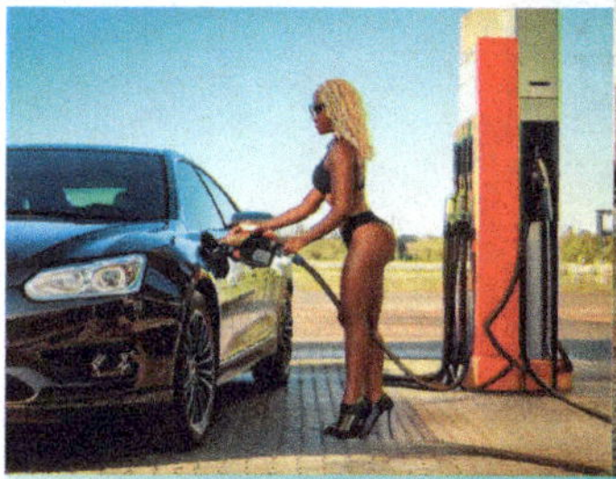

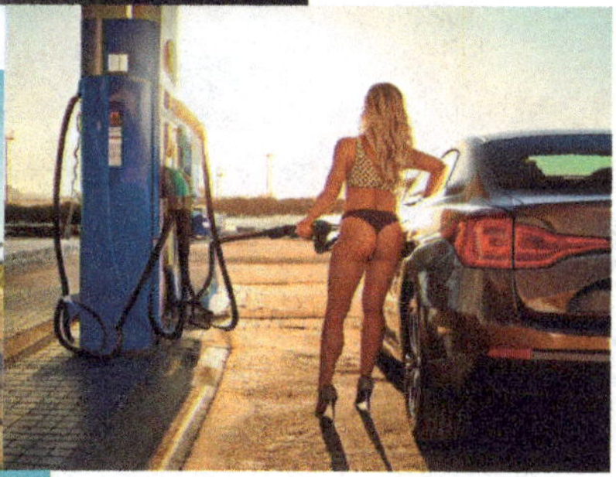

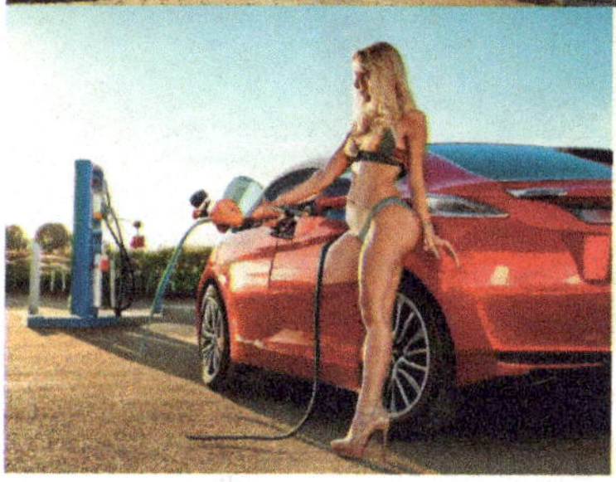

See distorted bellies and such.

full body shot of woman with blond hair in high heels and bathing suit pumping gas into sports car at gas stati

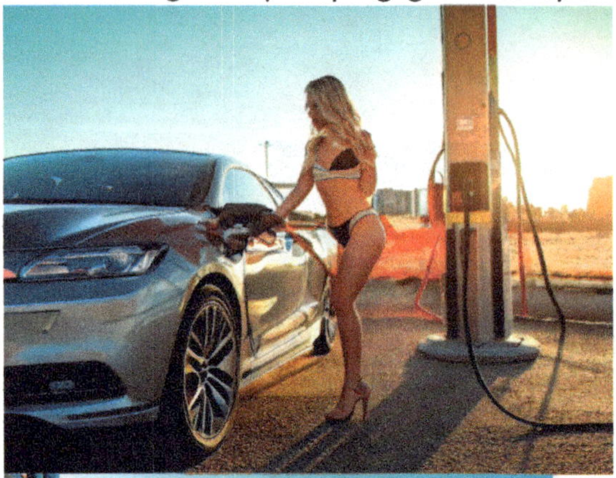

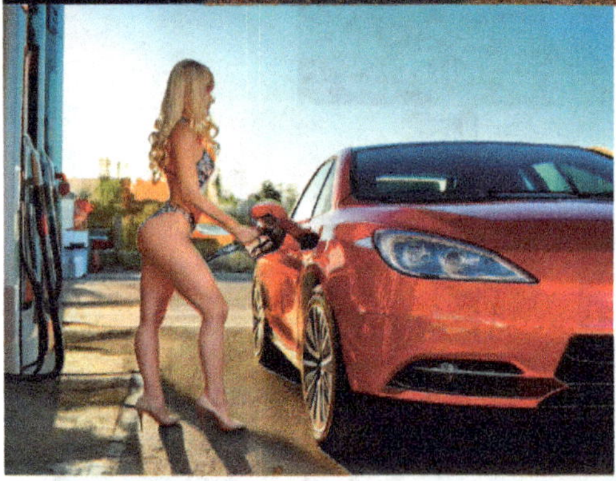

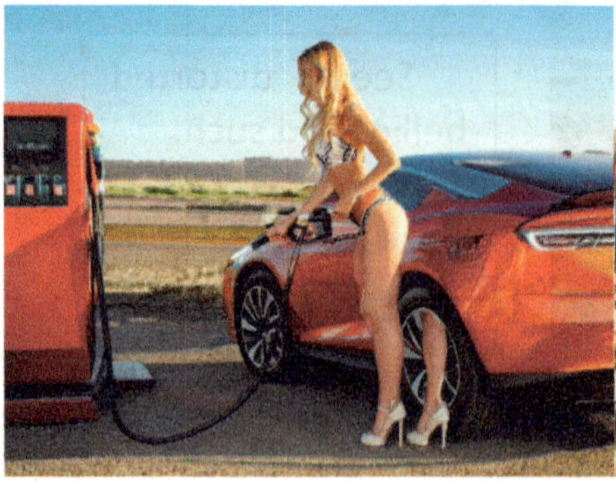

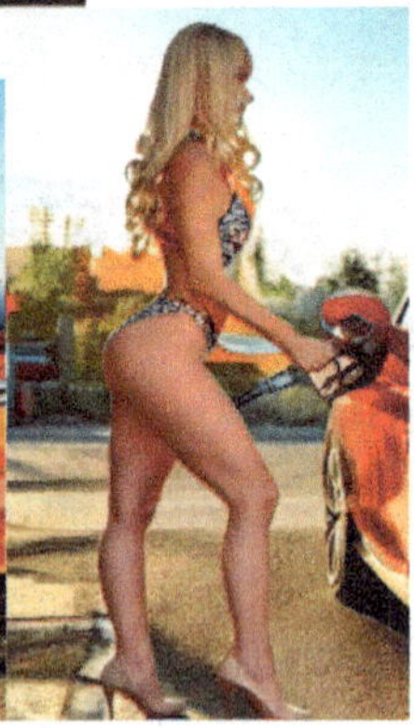

full body shot from above of woman with blond hair in high heels and bathing suit pumping gas into sports car at gas station

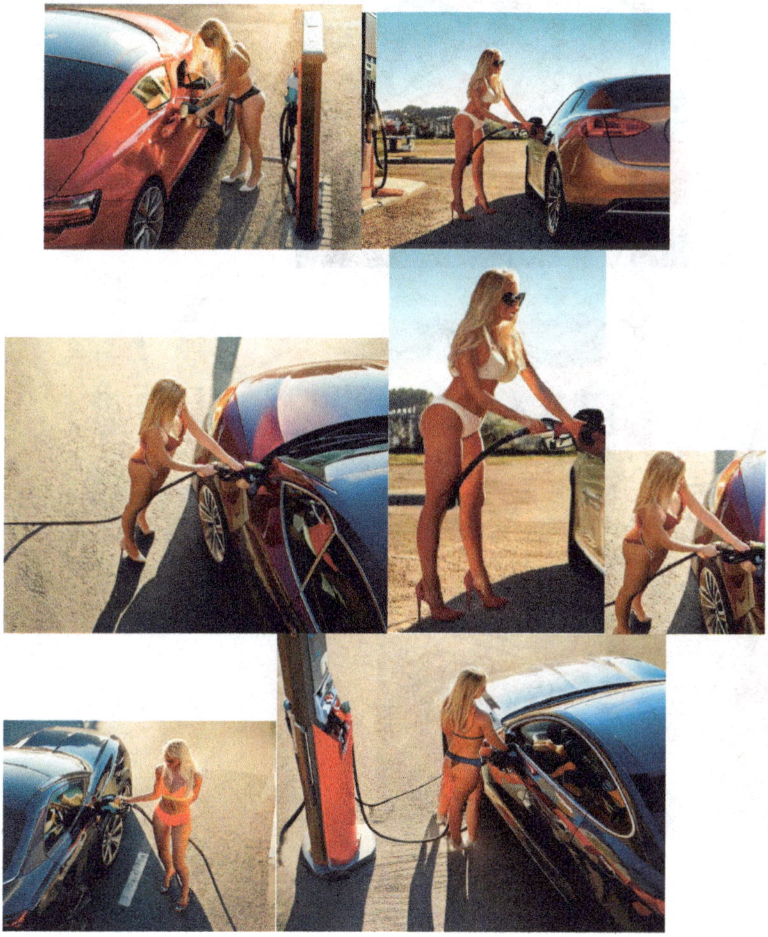

full body shot from above of woman with blond hair in high heels and bathing suit pumping gas into sports car at gas station

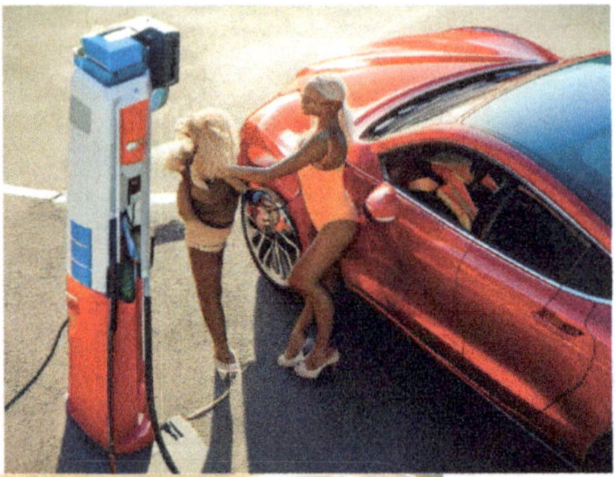

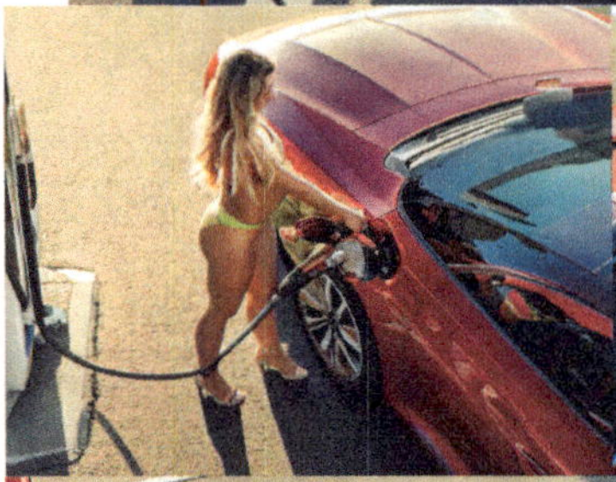

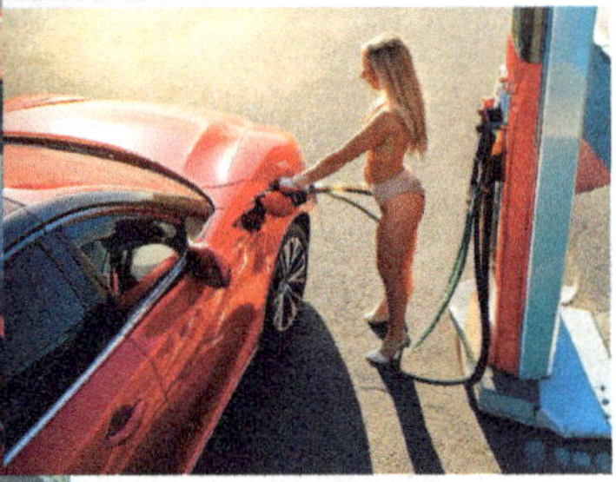

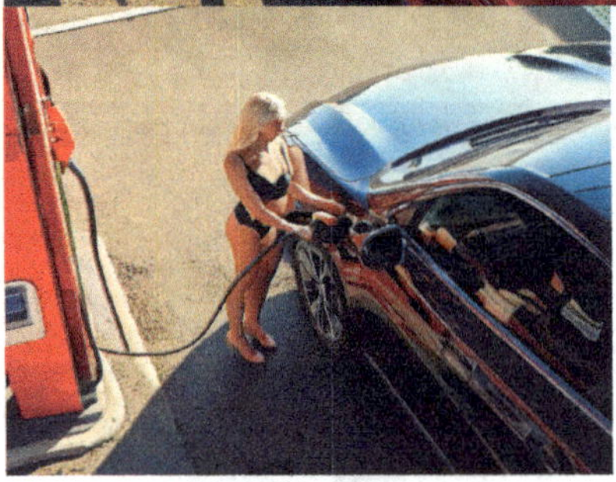

full body shot from above of woman in high heels and bathing suits pumping gas into sports car at gas station

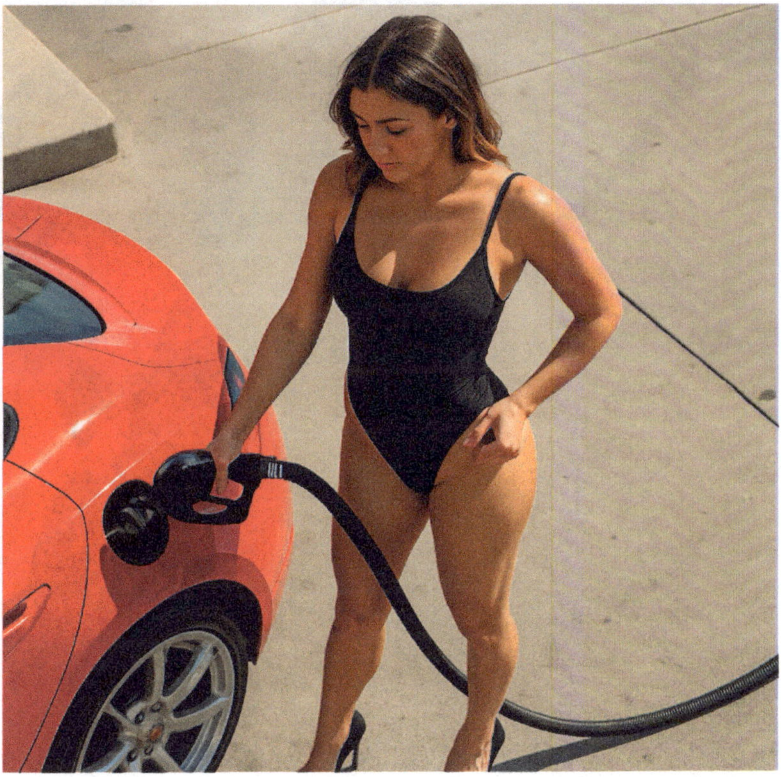

All of a sudden, the Adobe program started giving me shots more in line with what I had expected to be able to ask for. Boy, the program sure likes black onesy bathing suits. Did I mention that in the begining , it wouldn't do the word "bikini" but, later, I got "string bikini bathing suit" to work.

full body shot from above of woman in high heels and bathing suits pumping gas into sports car at gas station

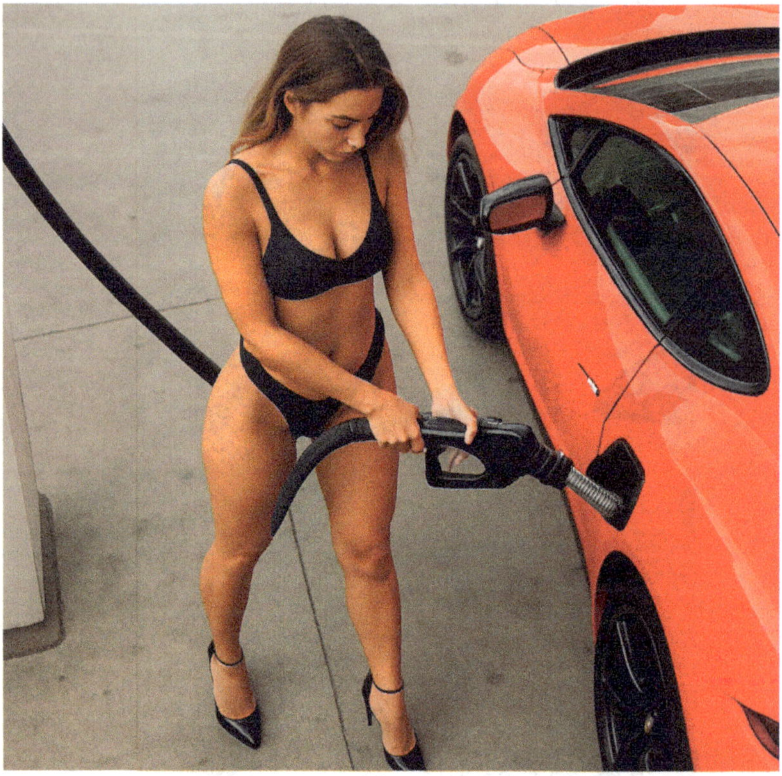

full body shot from above of woman in high heels and bathing suits pumping gas into sports car at gas station

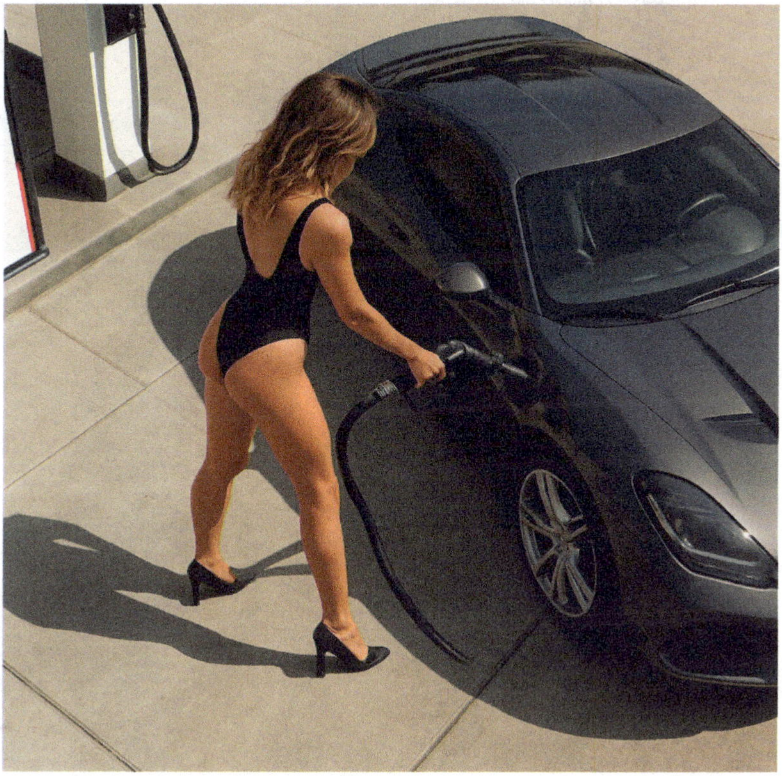

What's up with the robotic poses?

full body shot from above of woman in high heels and bathing suits pumping gas into sports car at gas station

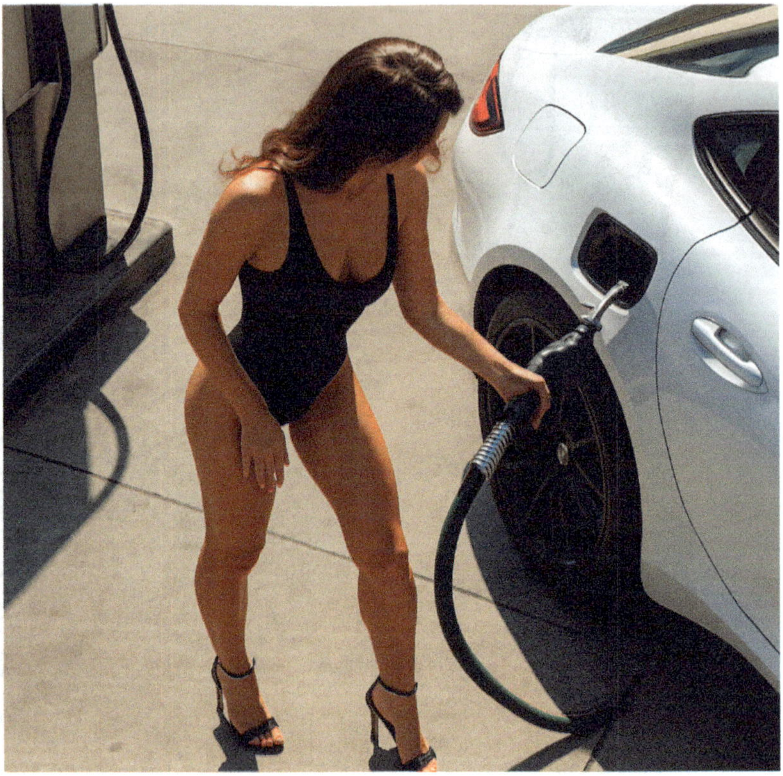

full body shot from above of woman in high heels and bathing suits pumping gas into sports car at gas station

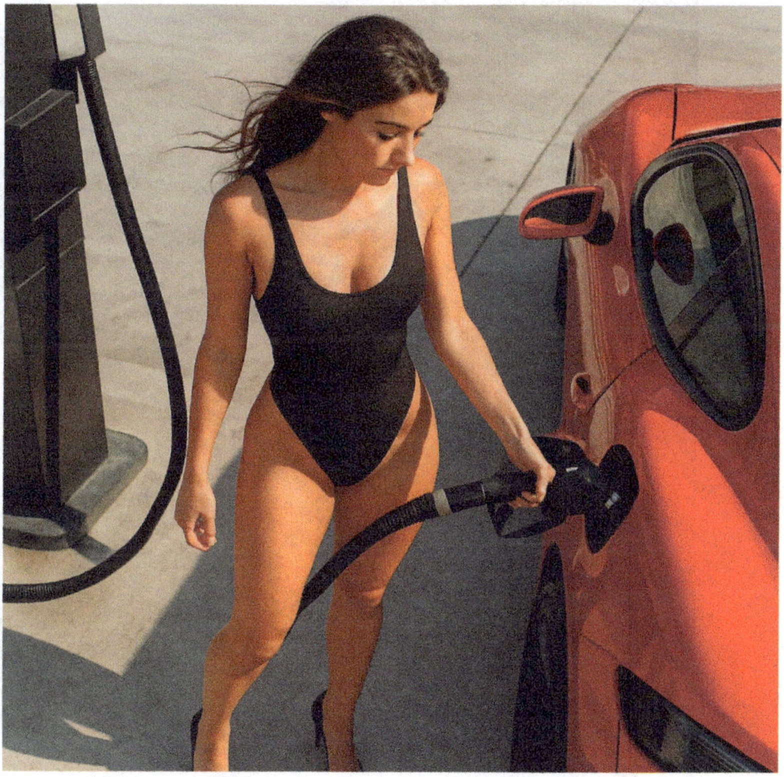

full body shot from above of woman in high heels and bathing suits pumping gas into sports car at gas station

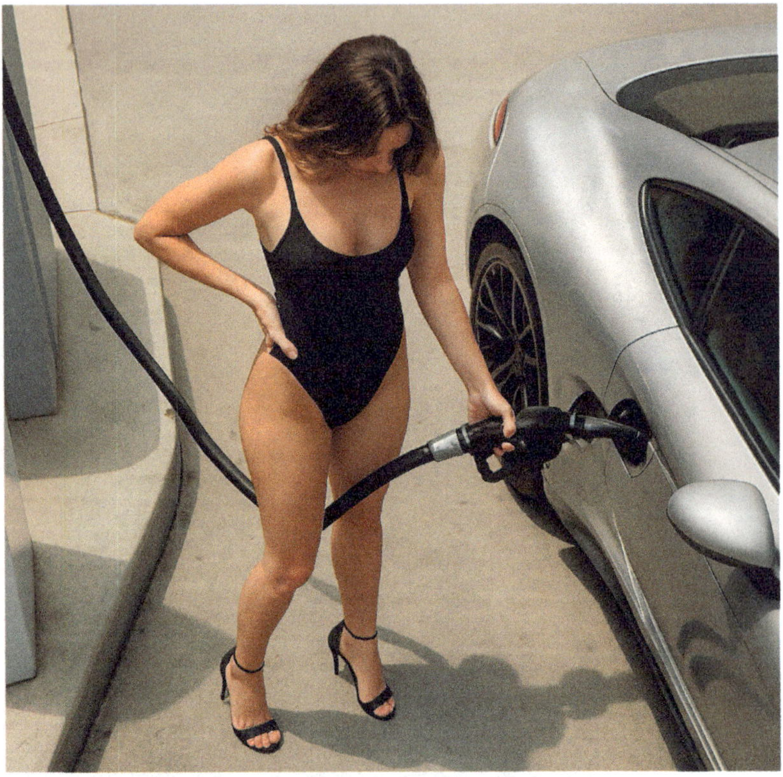

full body shot from below of woman in high heels and bathing suits pumping gas into sports car at gas station

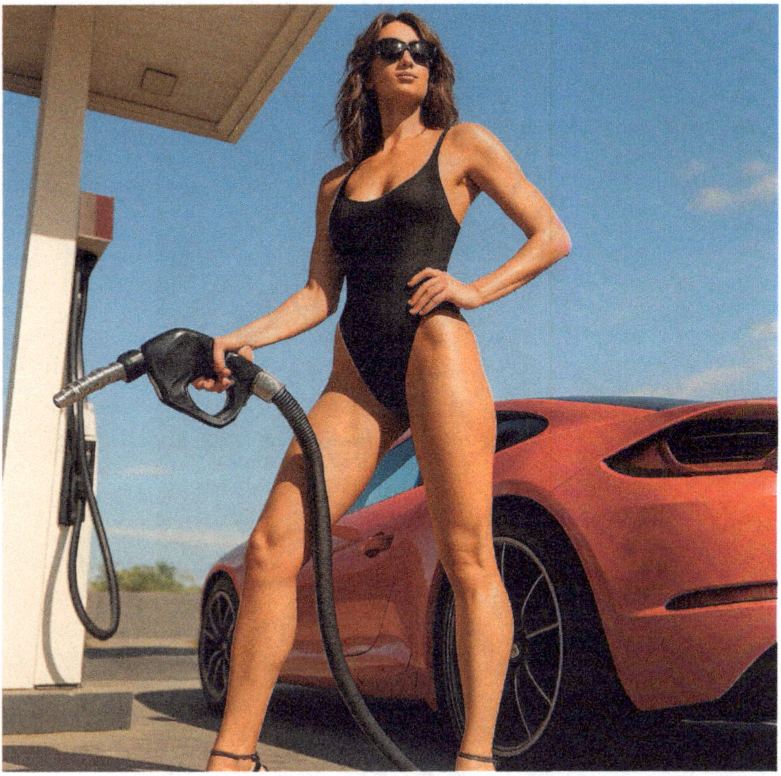

What's up with the hoses from nowhere?

full body shot from above of woman in high heels and bathing suits pumping gas into sports car at gas station

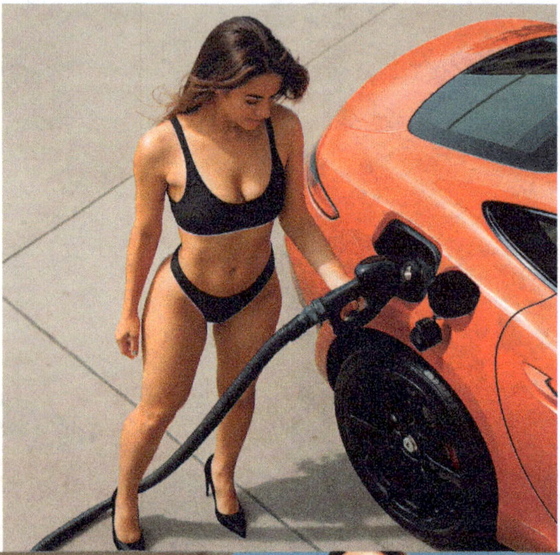

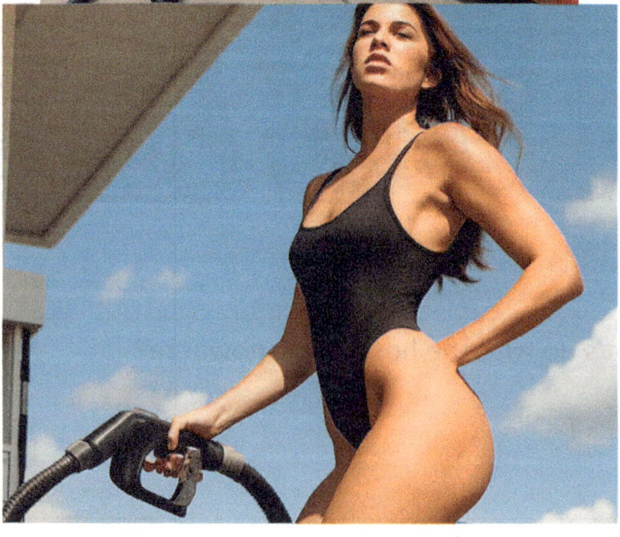

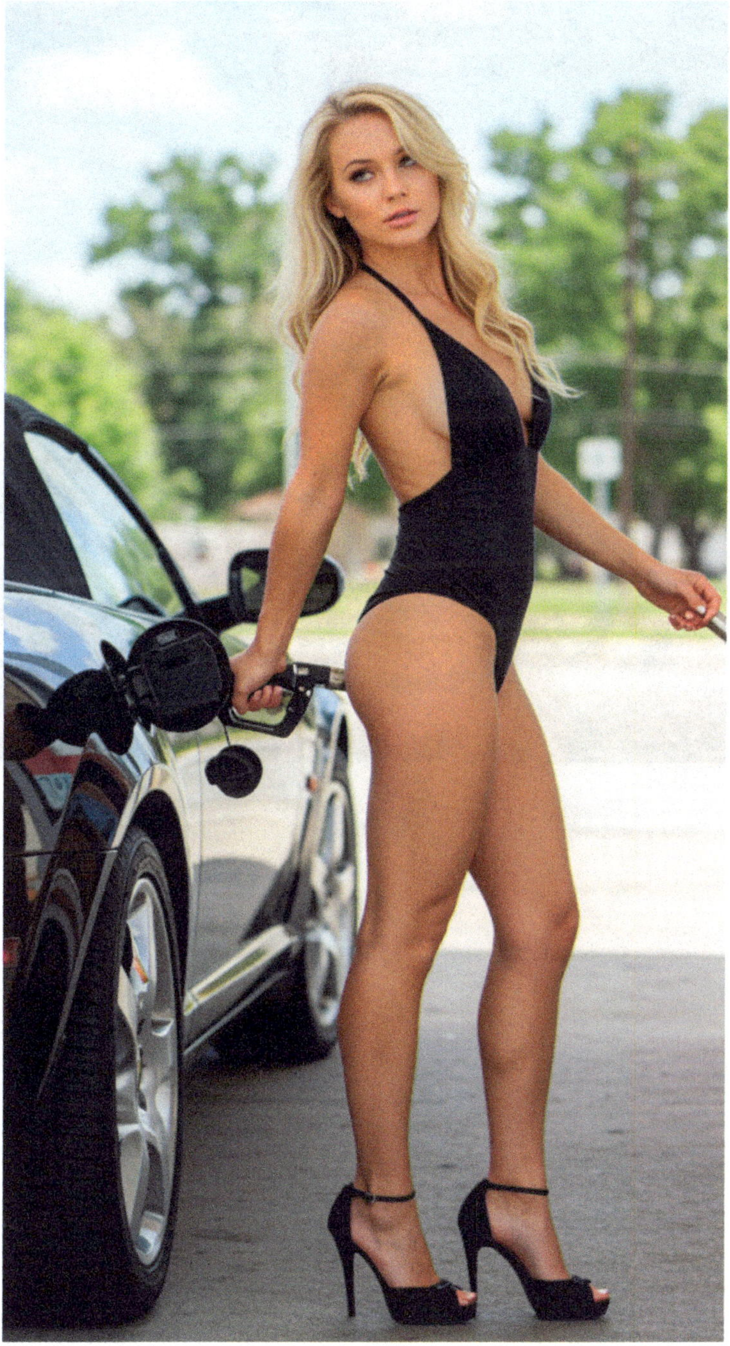

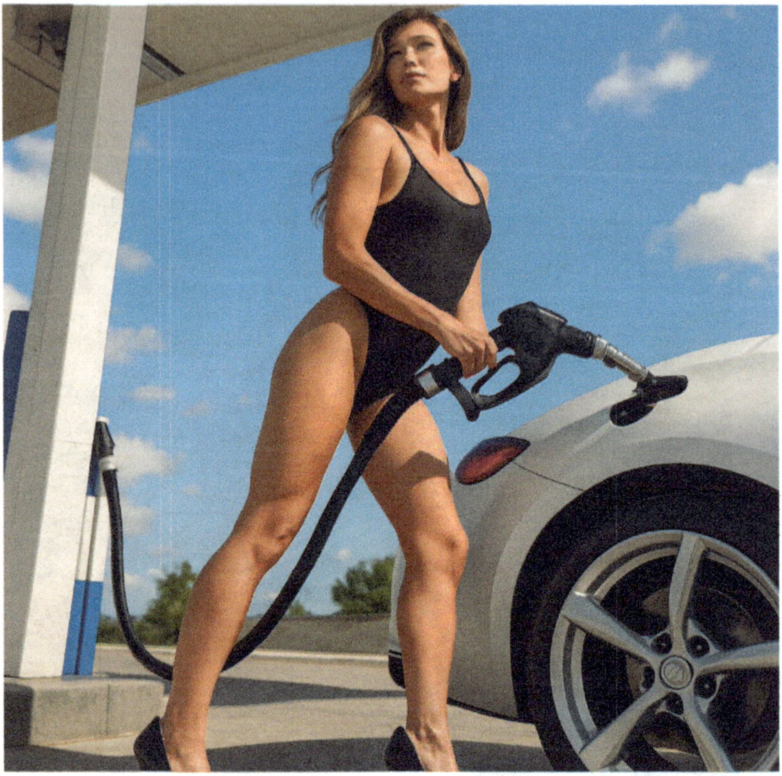

full body shot from below of blond woman in high heels and bathing suits pumping gas into sports car at gas station

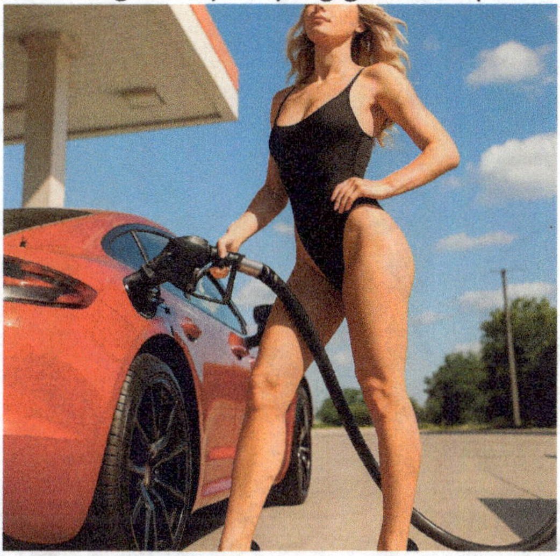

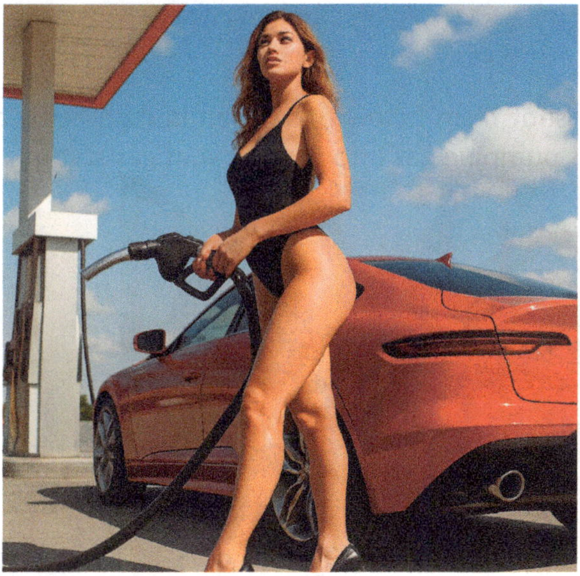

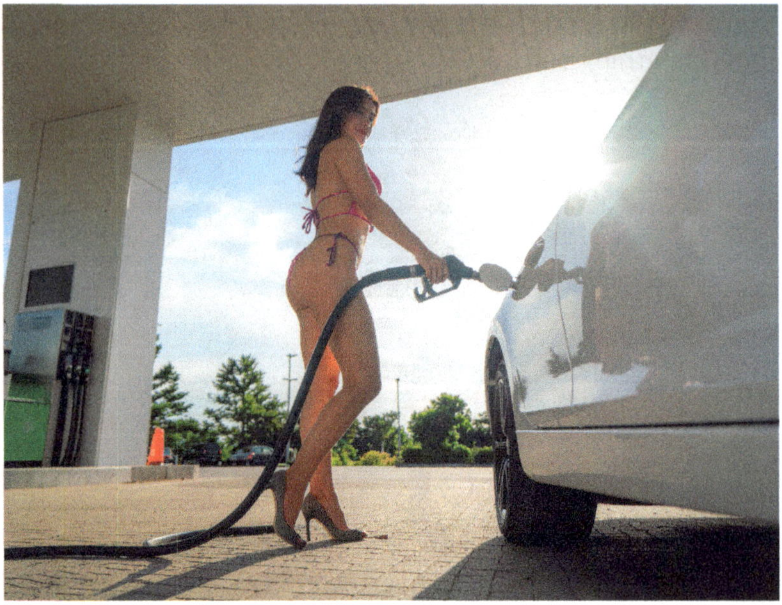

full body shot from below of woman in high heels and bathing suits pumping gas into sports car at gas station

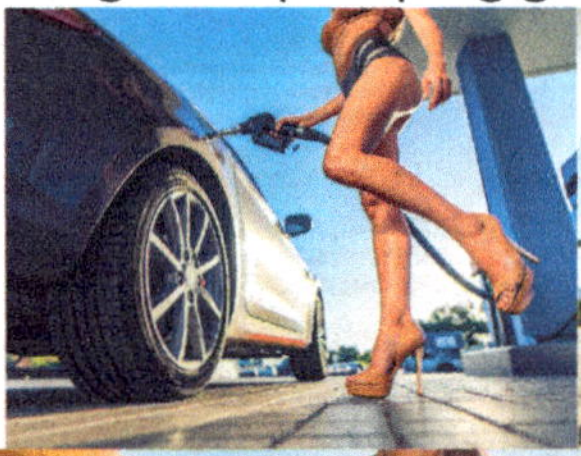
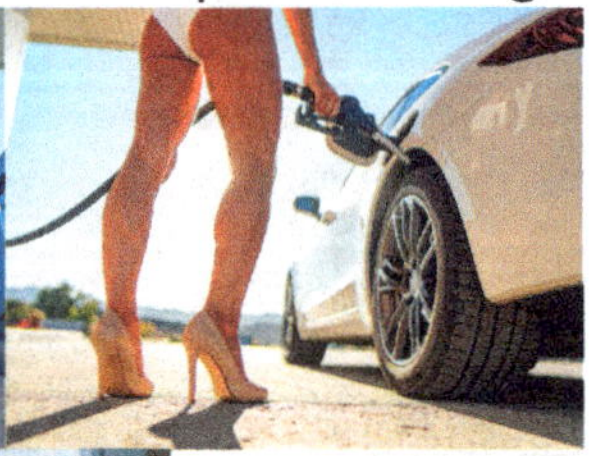
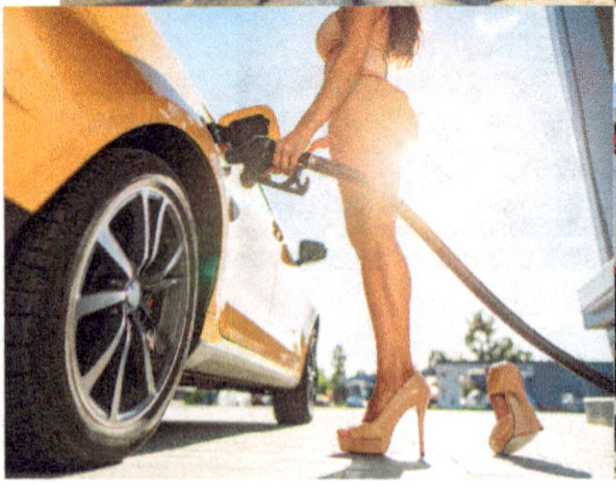
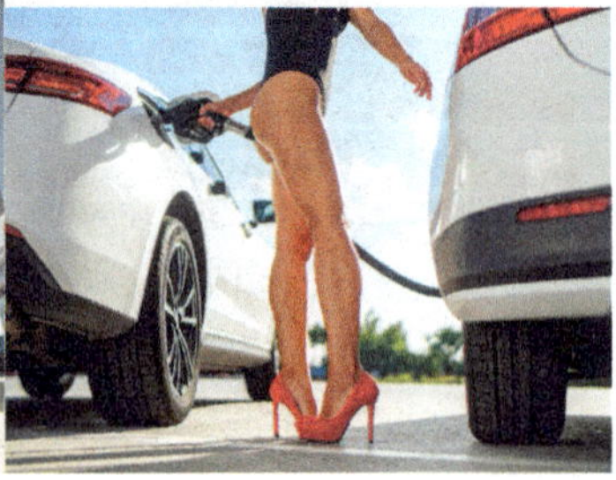

full body shot of woman in high heels and bathing suits pumping gas into sports car at gas station

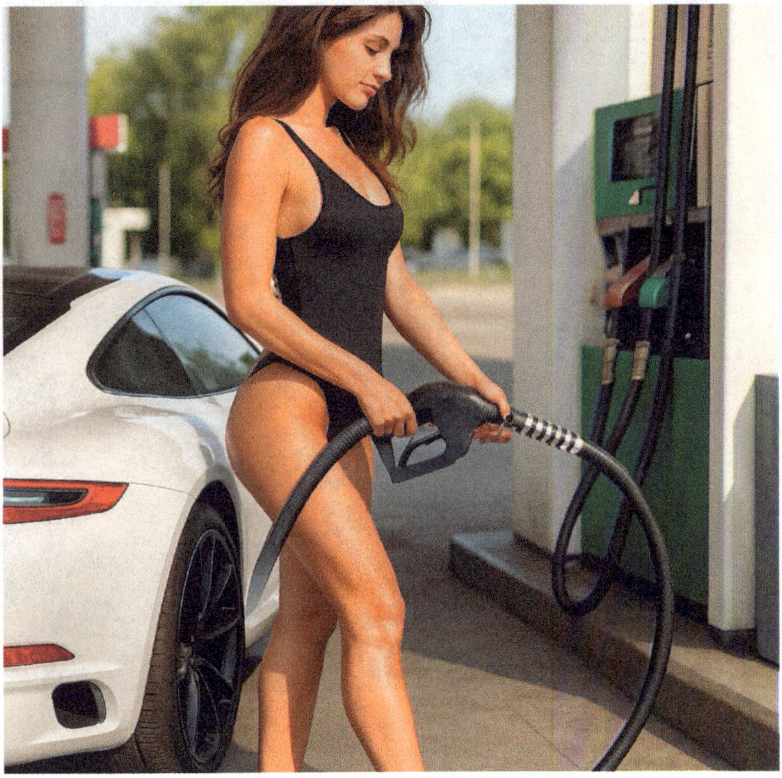

sexy woman in backless evening gown pumping gas at gas station

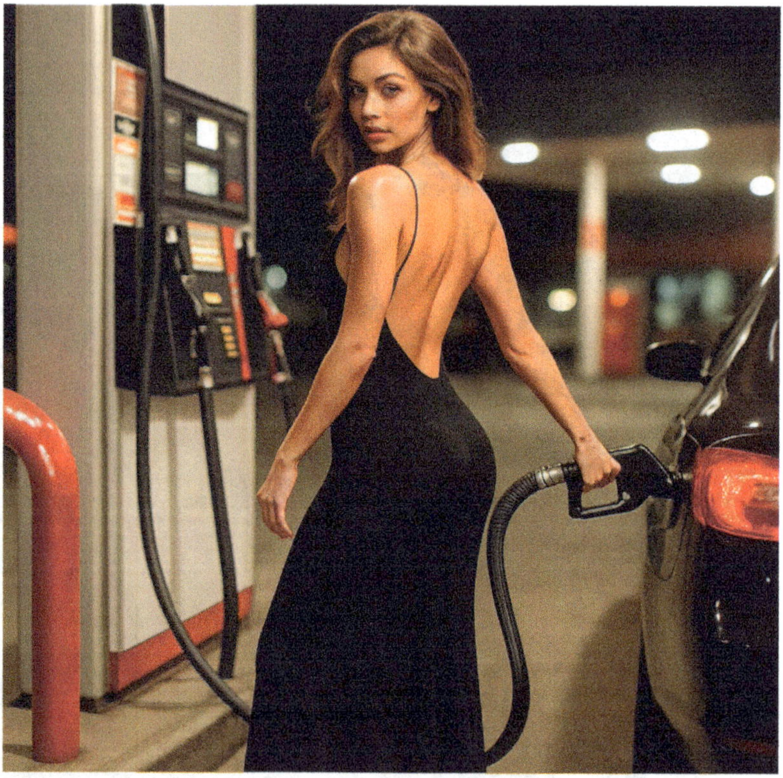

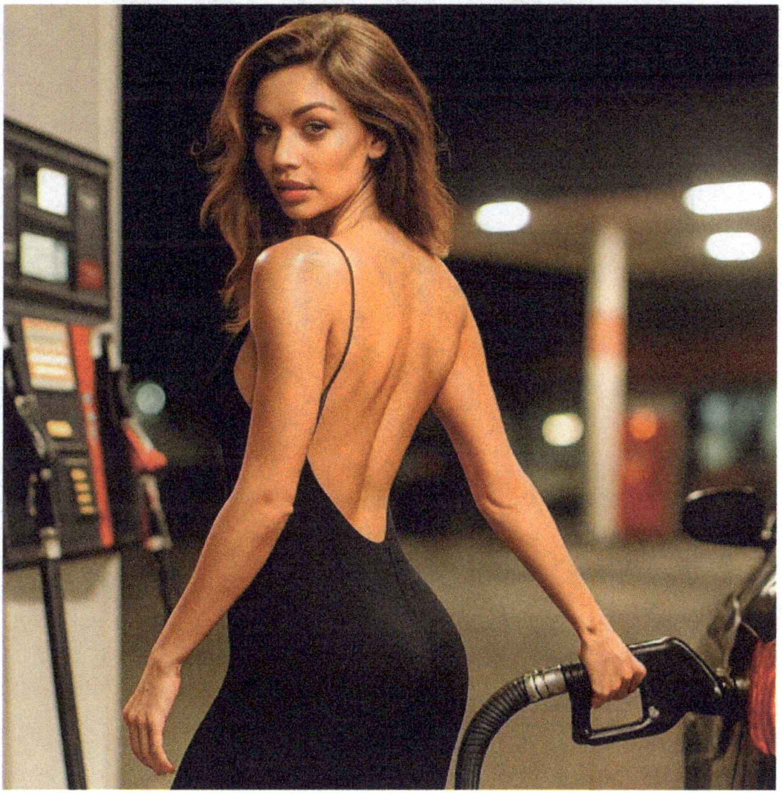

sexy woman in backless evening gown pumping gas at gas station

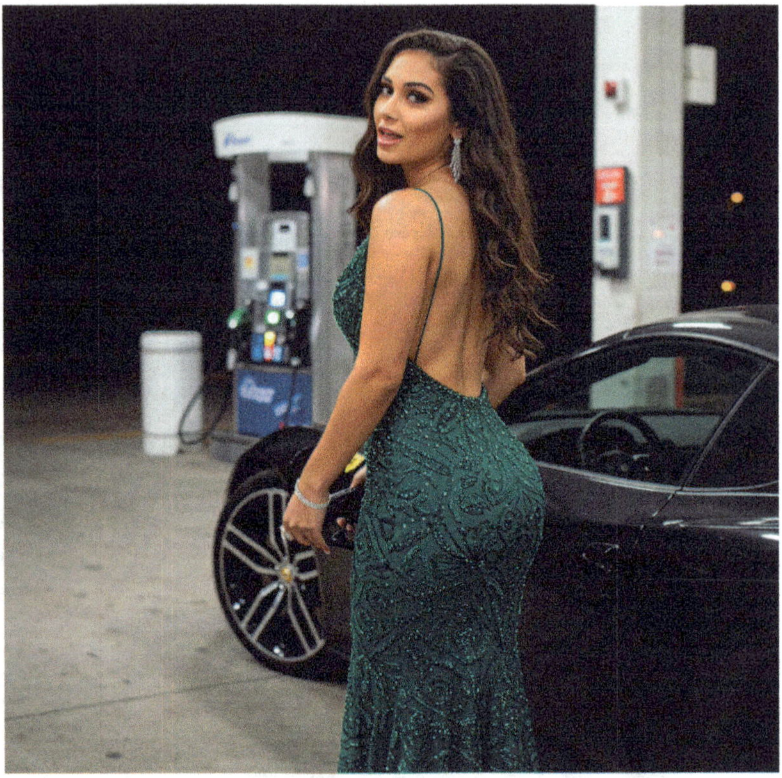

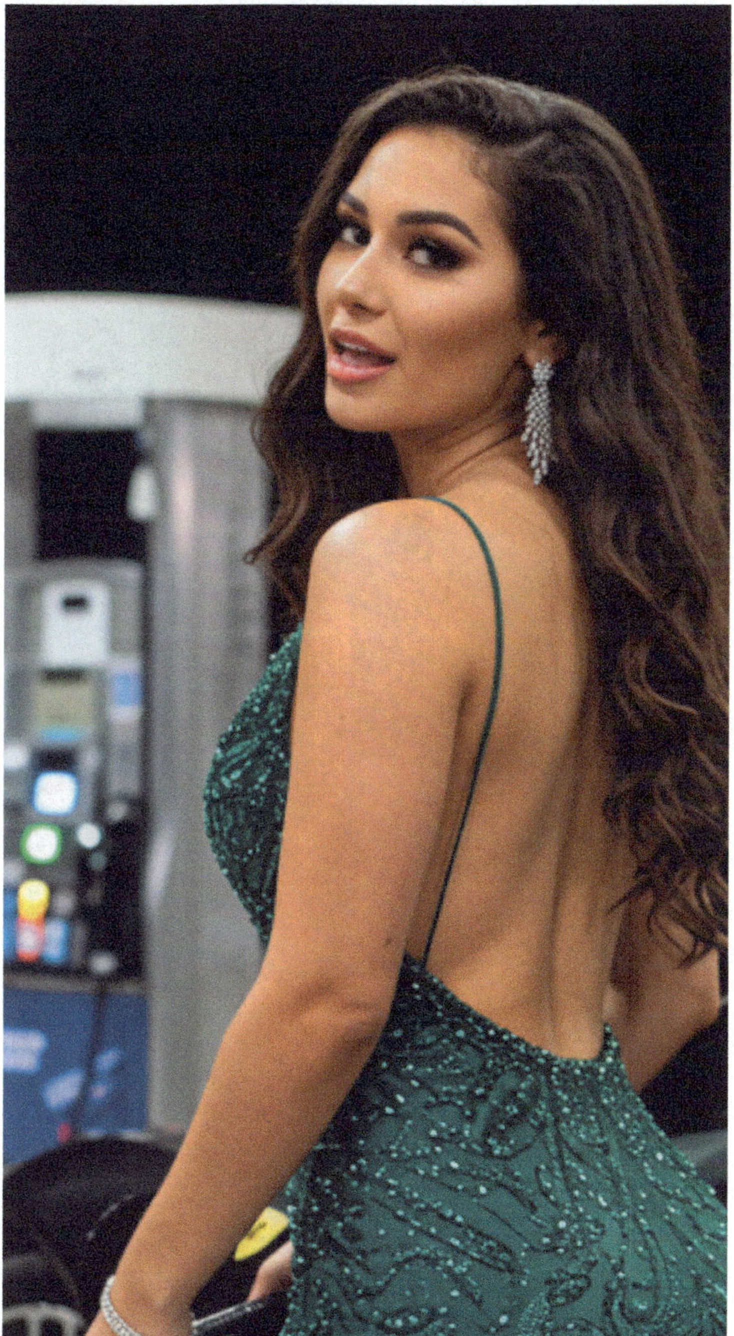

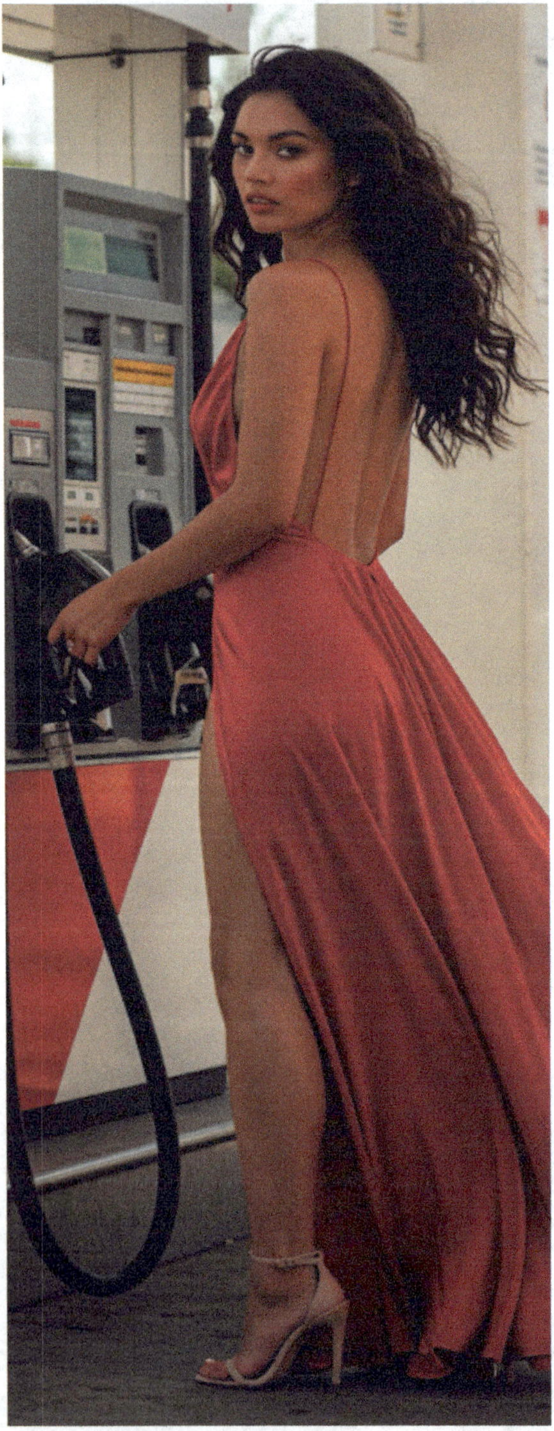

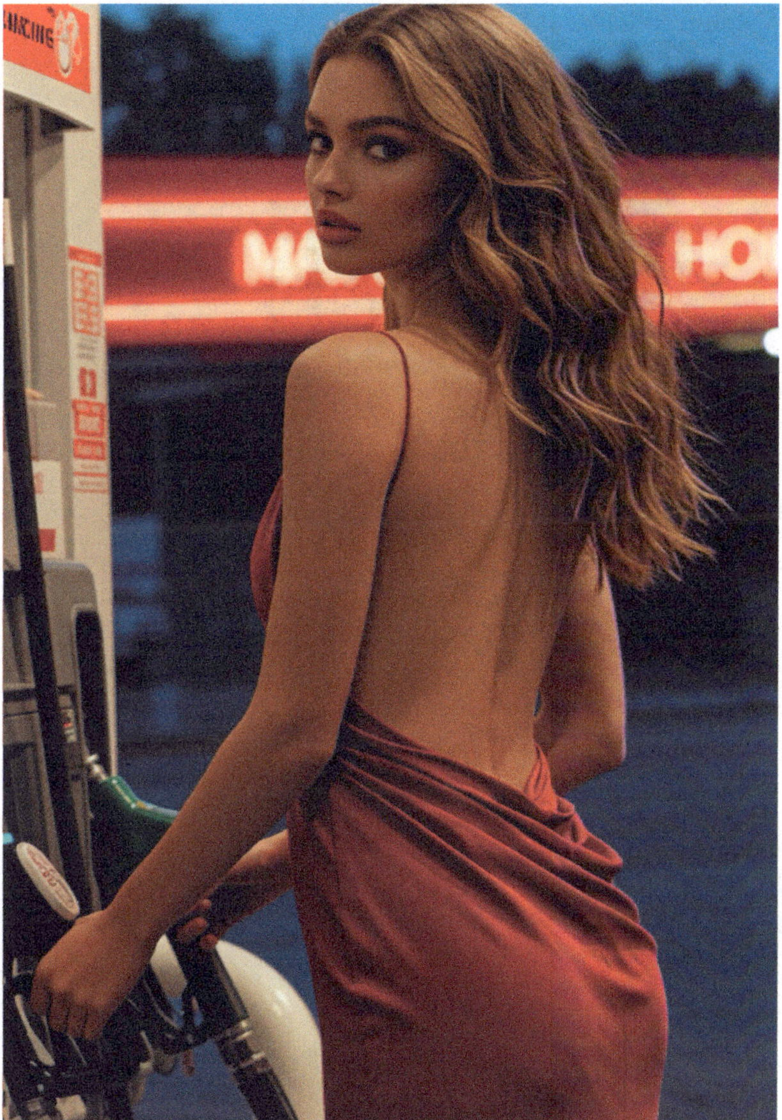

sexy red headed woman in backless evening gown pumping gas at gas station

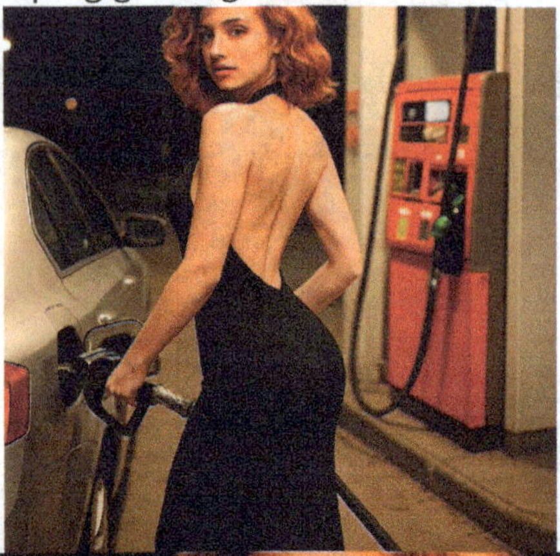

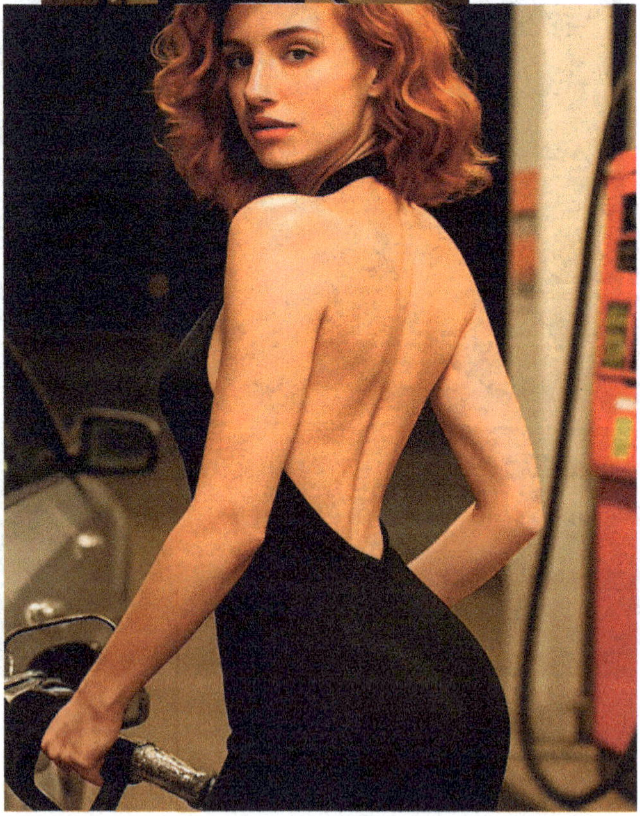

“Dressed for night club” seemed to inspire the program.

sexy woman dressed for night club pumping gas at gas station

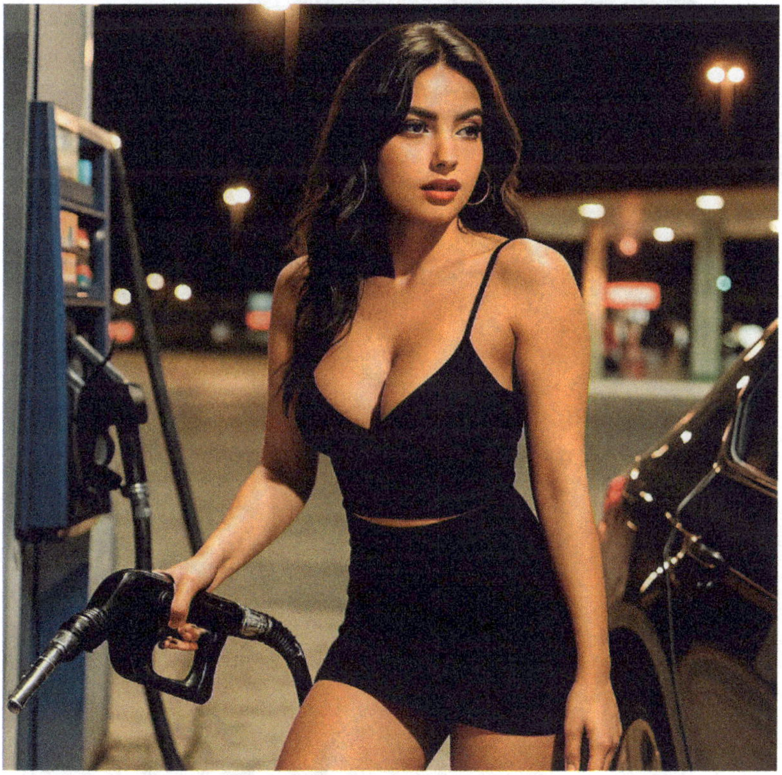

sexy woman dressed for night club pumping gas at gas station

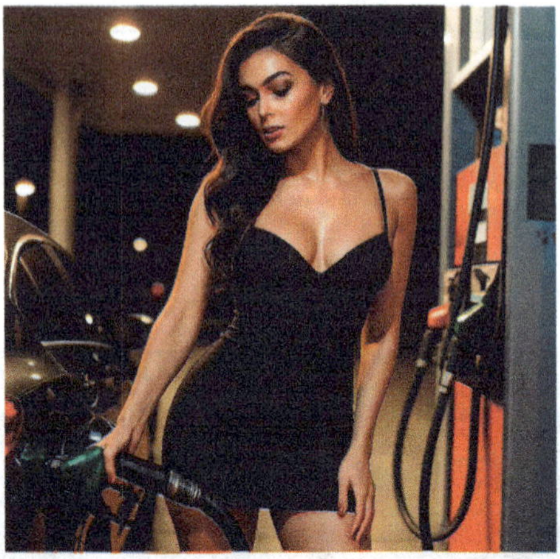

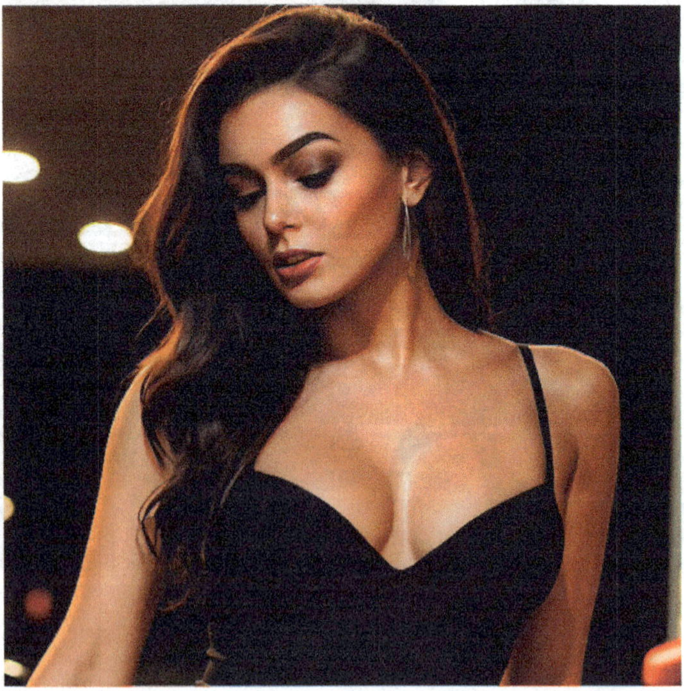

sexy blond woman dressed for night club pumping gas at gas station

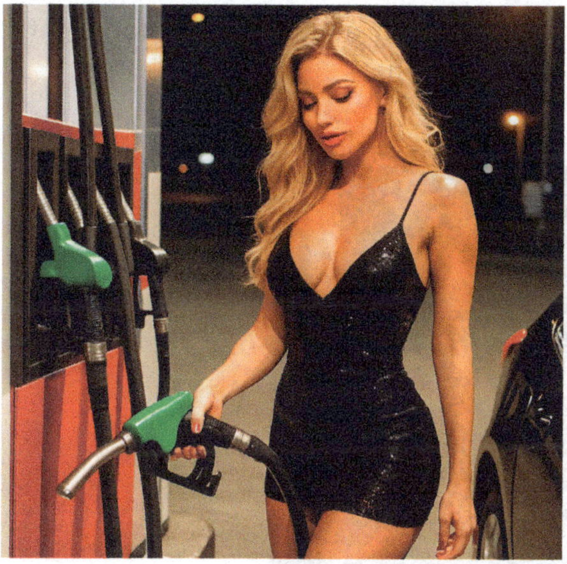

sexy woman dressed for night club pumping gas at gas station

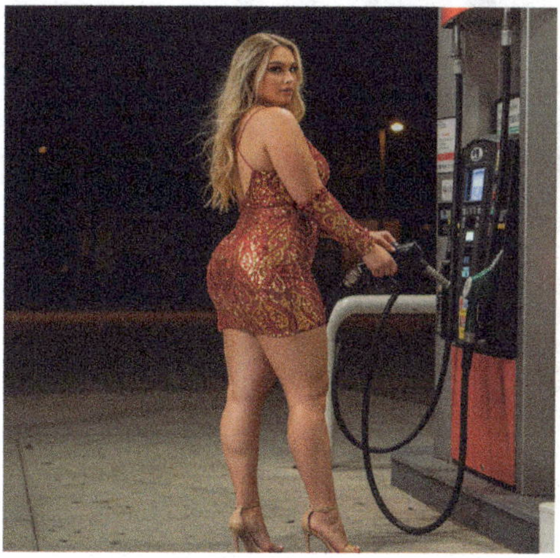

sexy woman shot from above dressed for night club pumping gas at gas station

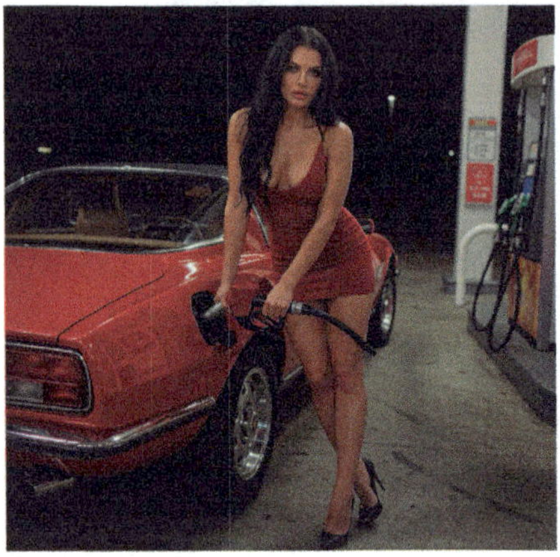

beautiful woman pumping gas

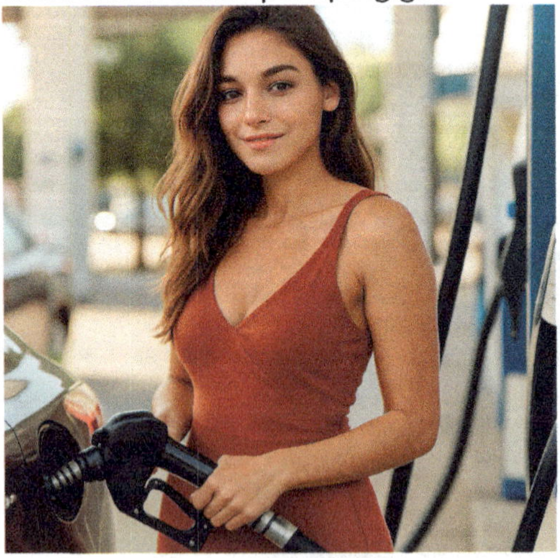

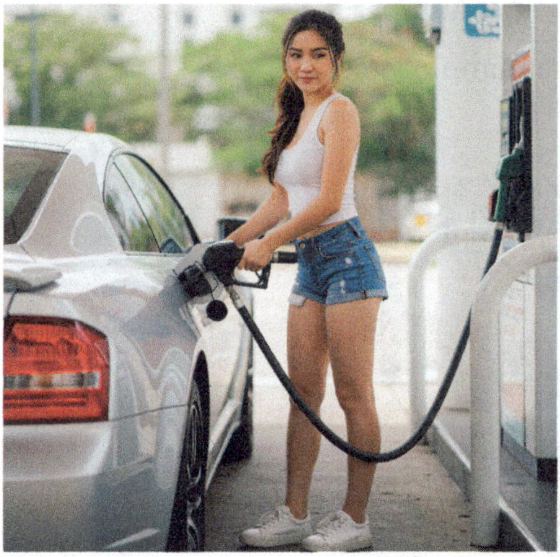

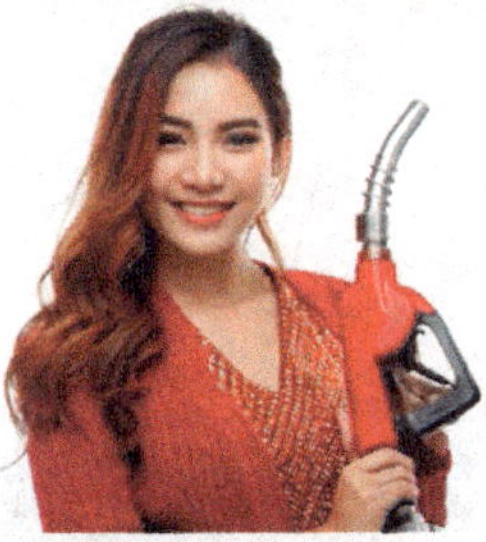

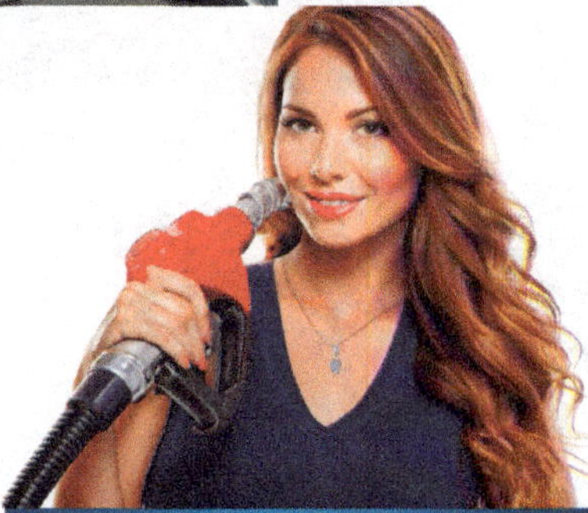

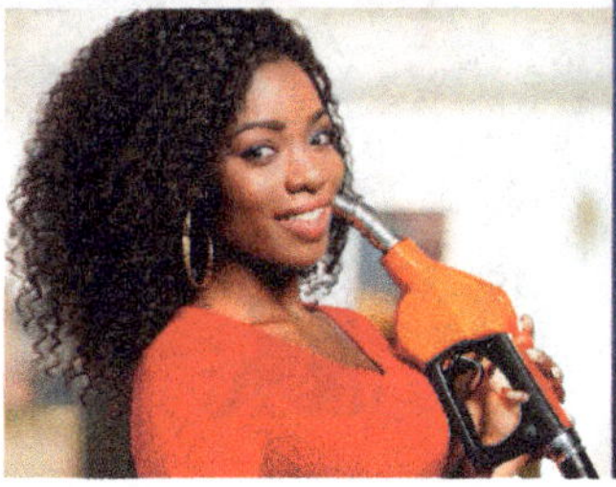

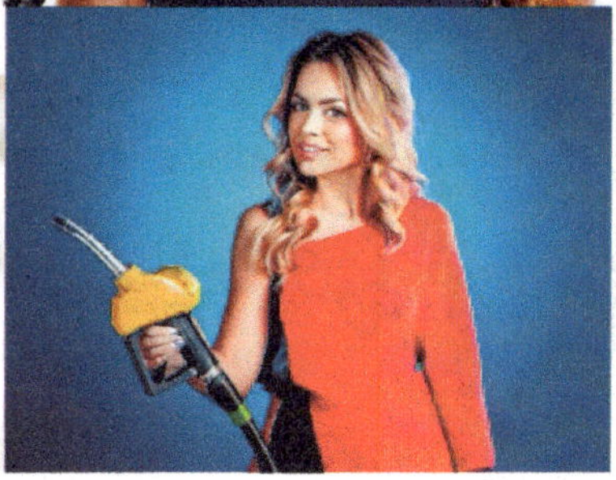

Again, notice differences in style of "stock image" type of pictures developed by different programs, but also, by different degrees of simplicity of the prompt.

beautiful woman pumping gas

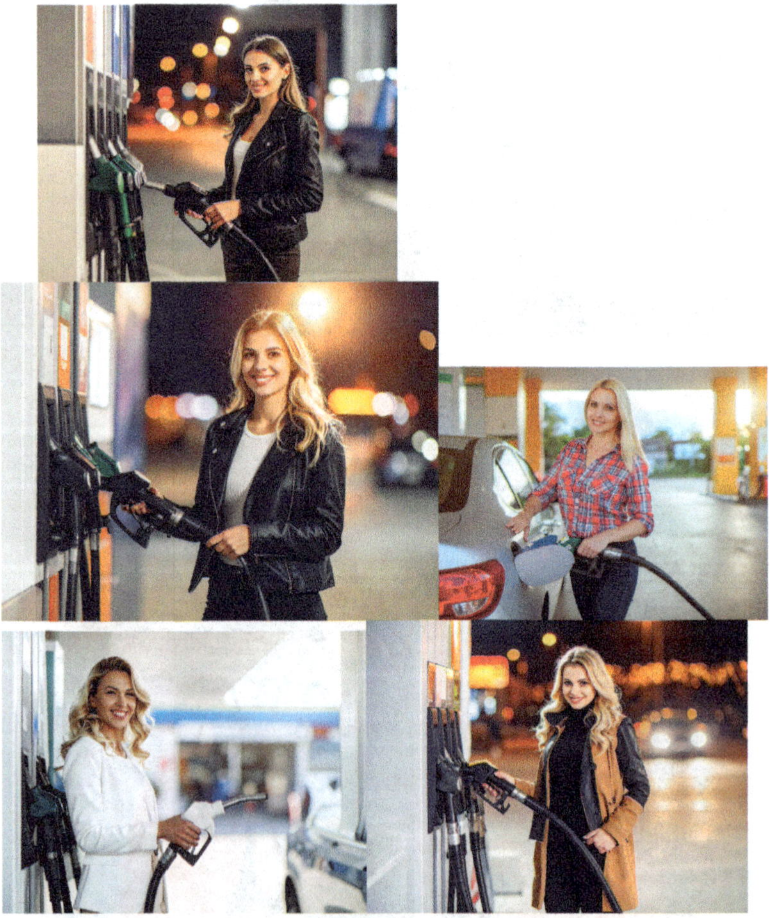

beautiful woman built like a brick shit house pumping gas

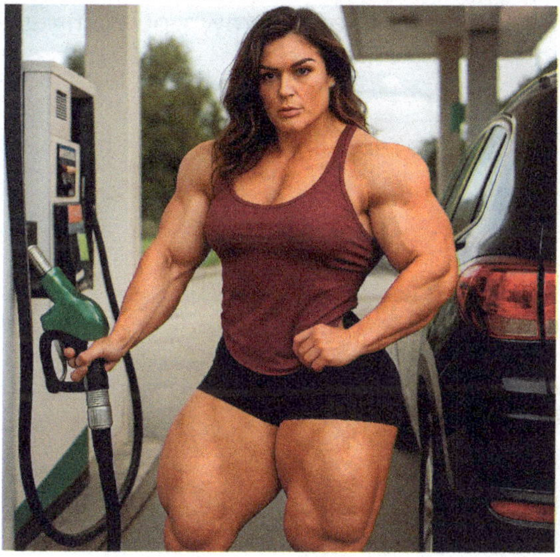

summer issue of vogue model pumping gasoline

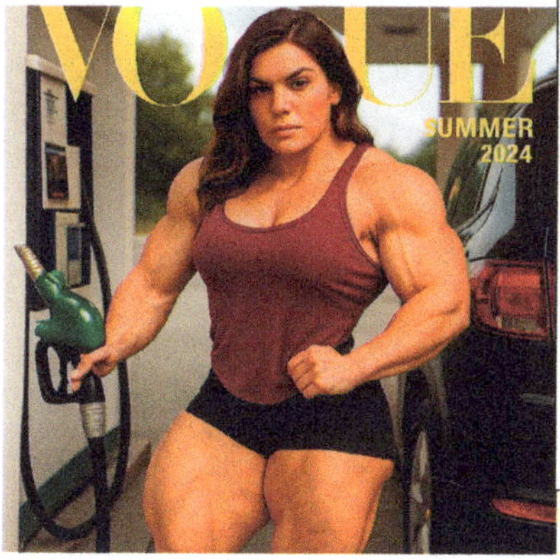

Here, "brick shithouse" model was regenerated even though I did the above prompt, not trying to use same model but program did similar image just because it was the next query..

sexy woman in backless evening gown pumping gas at gas station

summer issue of vogue model pumping gasoline

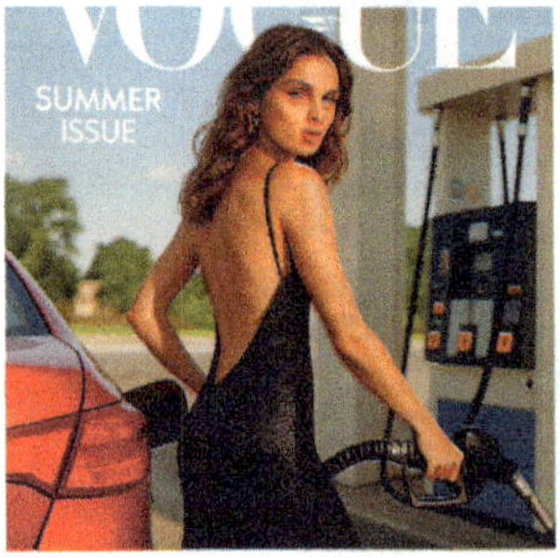

summer issue of vogue model pumping gasoline

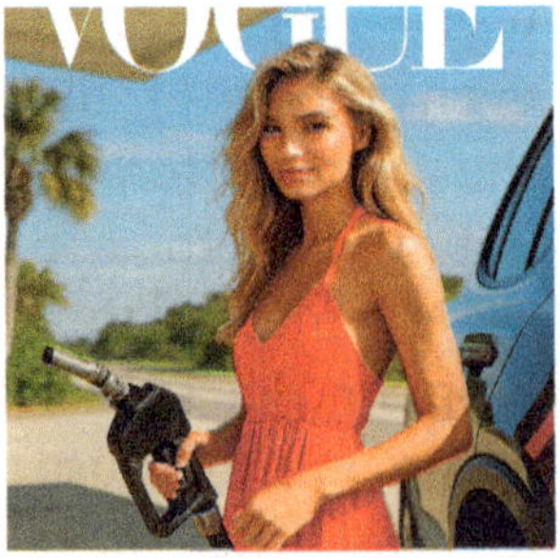

sexy woman in backless evening gown pumping gas at gas station summer issue of vogue model pumping gasoline

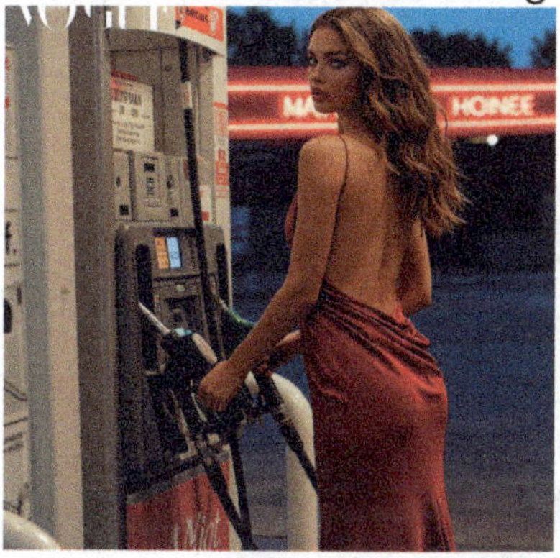

sexy blond woman in backless evening gown pumping gas at gas station summer issue of vogue model pumping gasoline

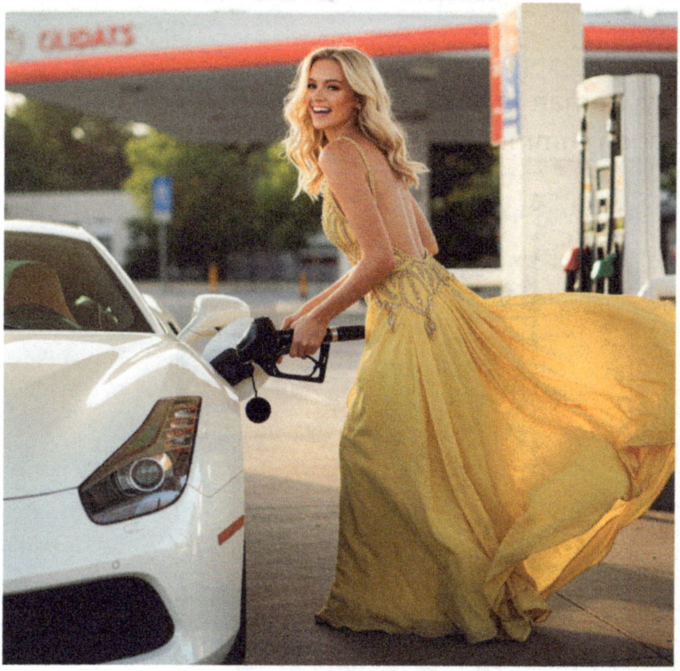

sexy blond woman in backless evening gown pumping gas at gas station summer issue of vogue model pumping gasoline

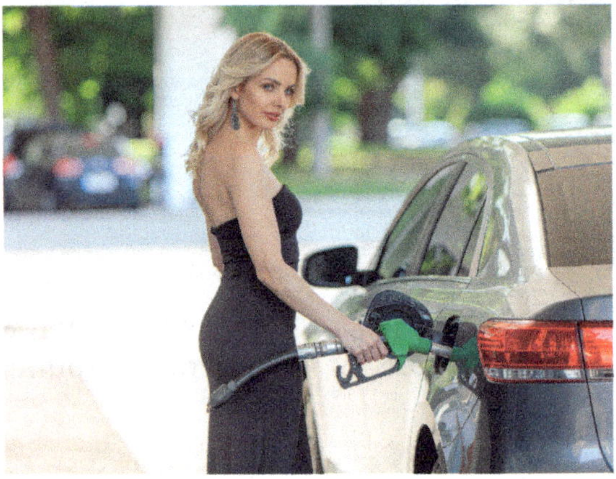

sexy woman in backless evening gown pumping gas at gas station summer issue of vogue model pumping gasoline

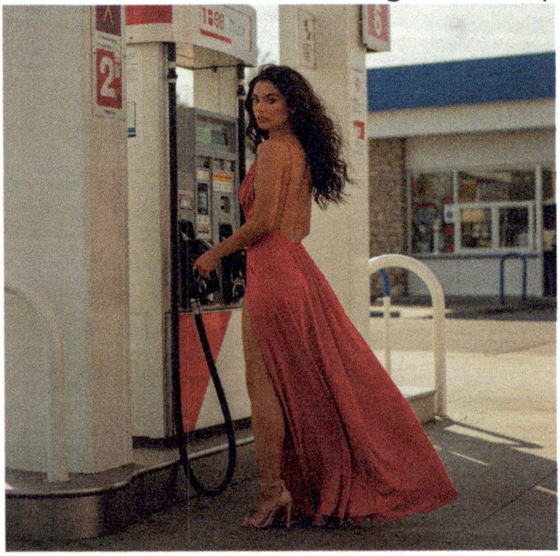

sexy woman in tube top and skinny jeans and high heels pumping gas at gas station summer issue of vogue model pumping g

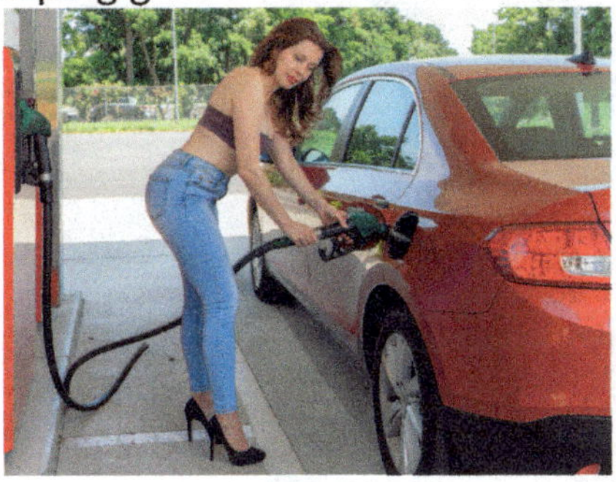

sexy woman in tube top and skinny jeans and high heels pumping gas at gas station summer issue of vogue model pumping gasoline

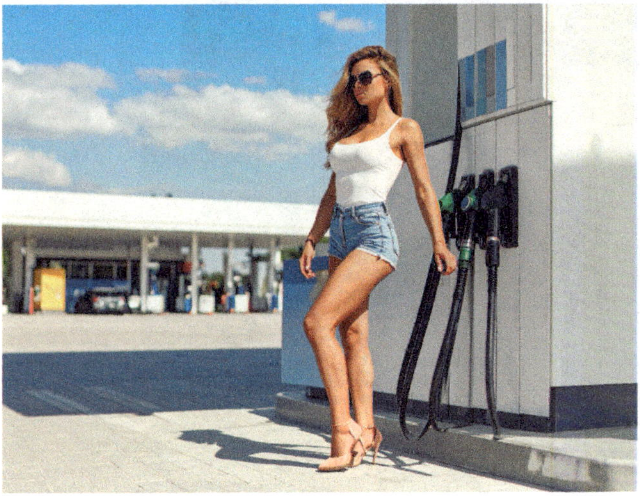

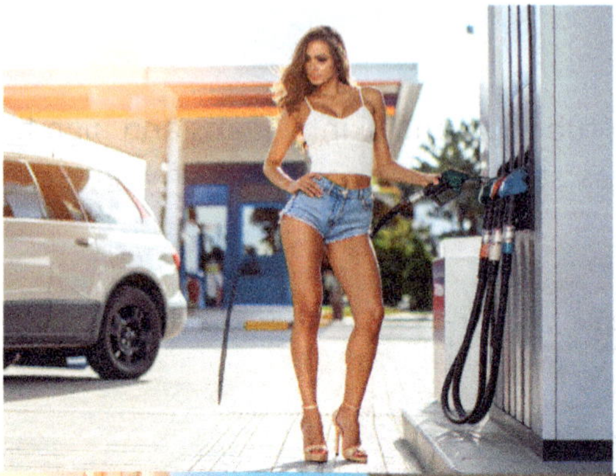
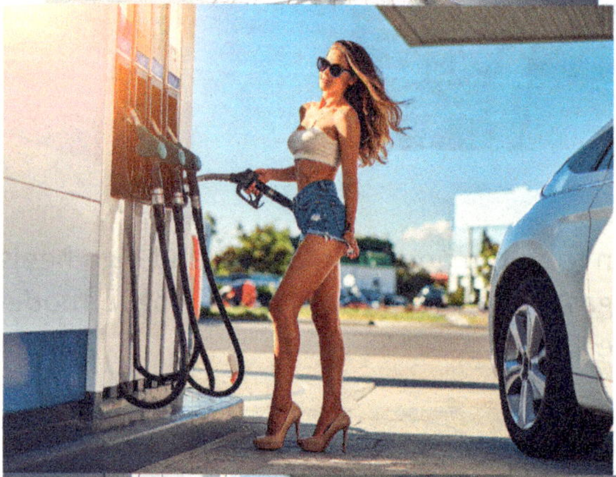
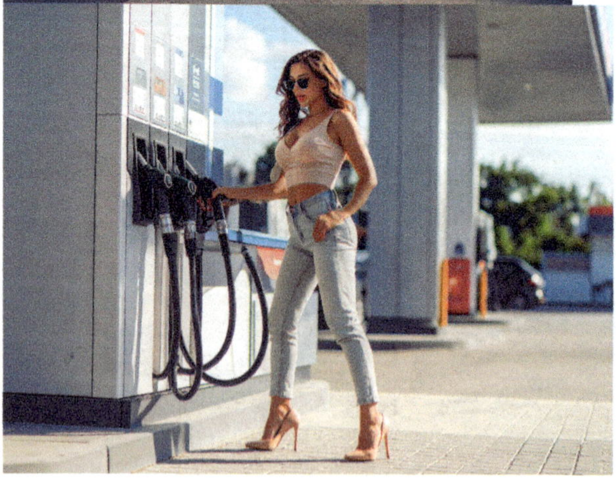

sexy woman in tube top and skinny jeans and high heels pumping gas at gas station summer issue of vogue model pumping gasoline

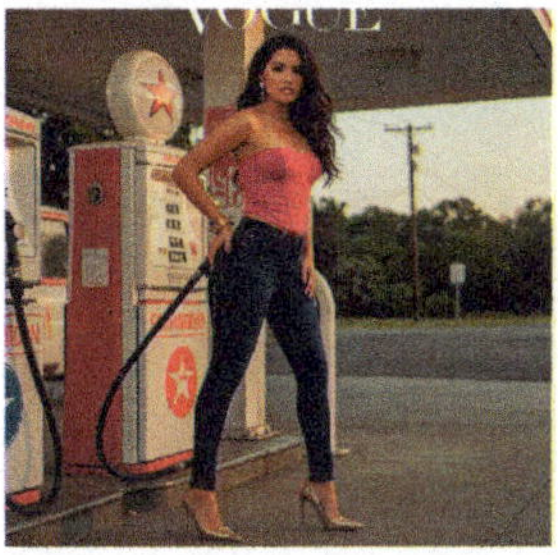

sexy woman in tube top and skinny jeans and high heels pumping gas at gas station summer issue of vogue model pumping gasoline

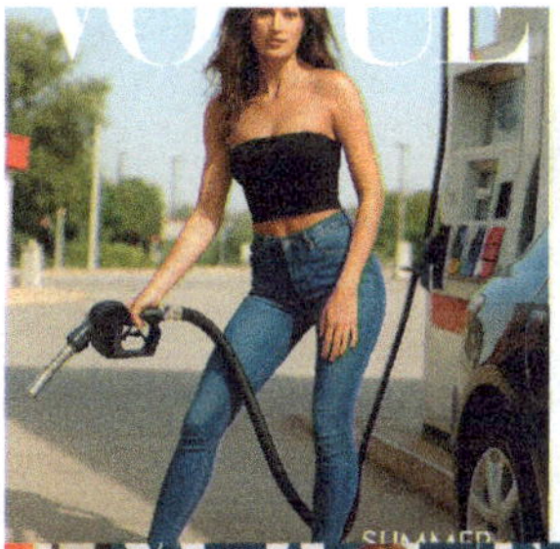

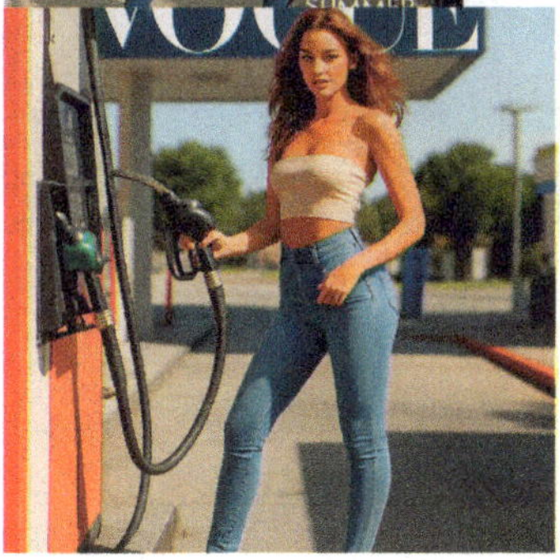

sexy woman in tube top and skinny jeans and high heels pumping gas at gas station summer issue of vogue model pumping gasoline

shot from above, a full body shot of sexy woman in tube top and skinny jeans and high heels pumping gas at gas station

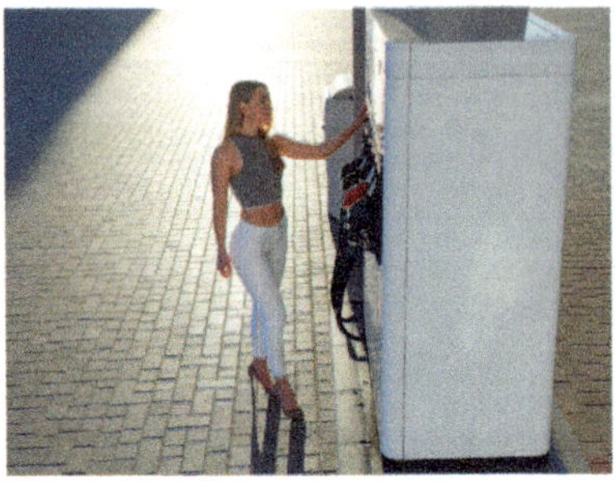

shot from above, a full body shot of sexy woman in tube top and skinny jeans and high heels pumping gas at gas station

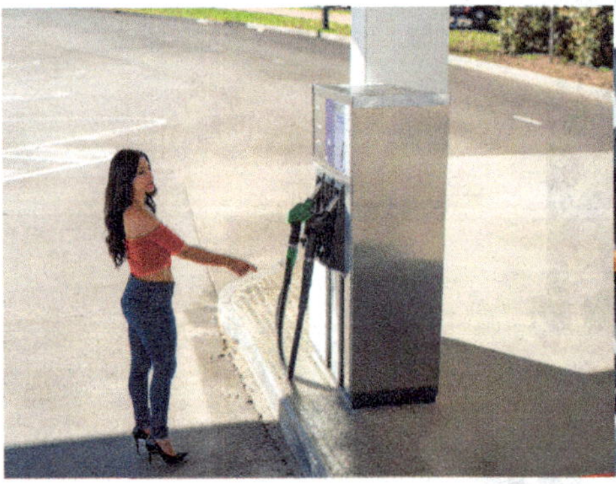

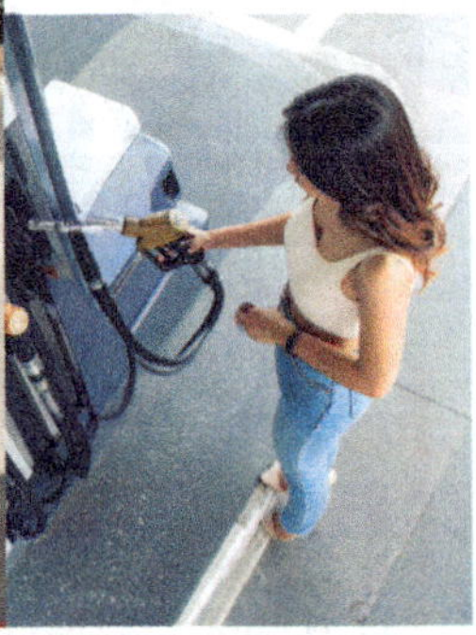

shot from above, a full body shot of sexy woman in tube top and skinny jeans and high heels pumping gas at gas station

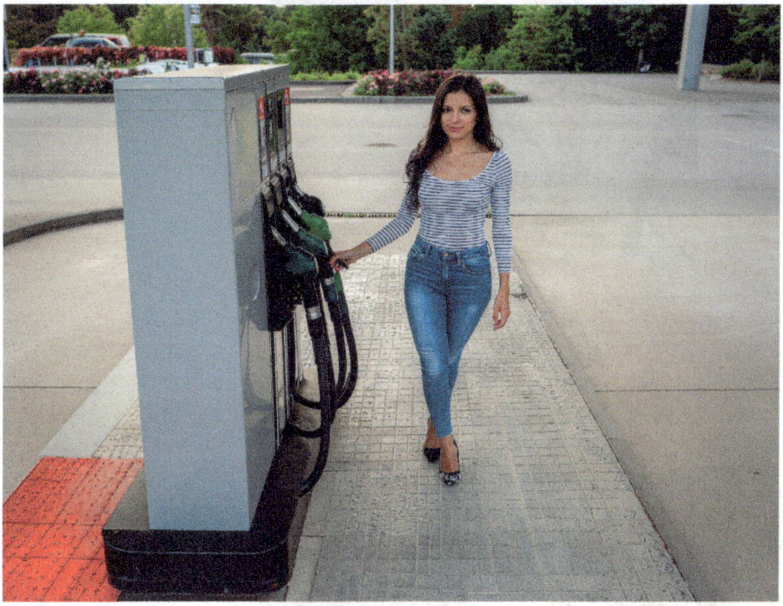

shot from above, a full body shot of sexy woman in tube top and skinny jeans and high heels pumping gas at gas station into fancy sports car

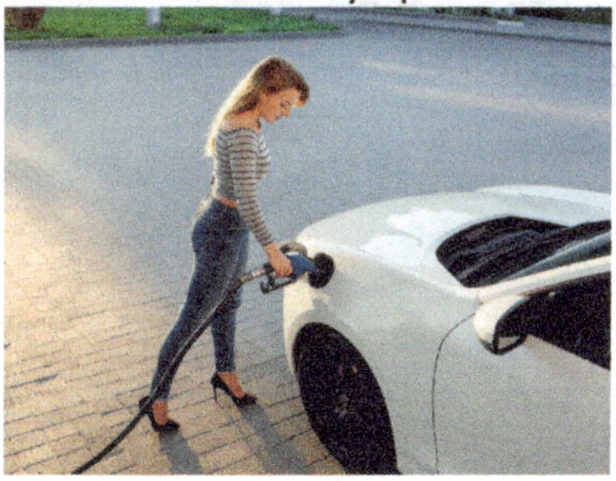

Strange how monochromatic the color palette became for a while.

shot from above, a full body shot of sexy woman in tube top and skinny jean shorts and high heels pumping gas at gas station into fancy sports car

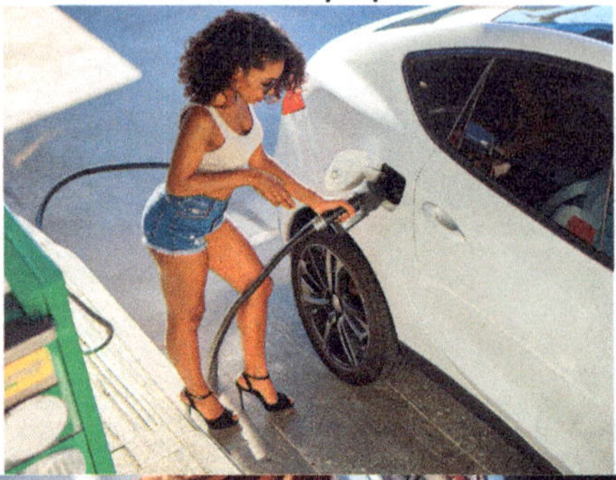

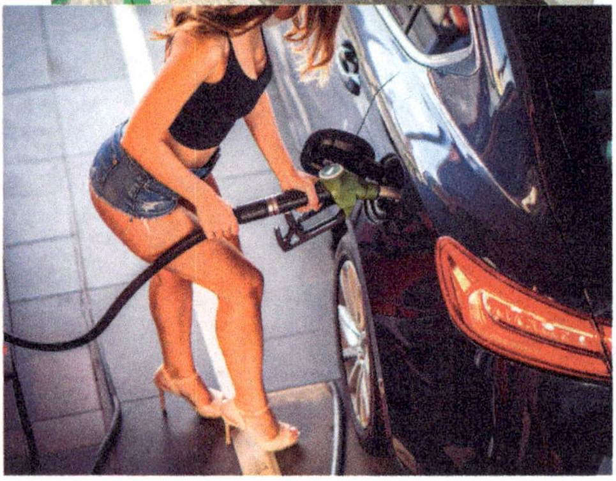

Fun to notice that even though the words seem to indicate where the focus of the image should be, sometimes the images get muddied and not focused on subject..

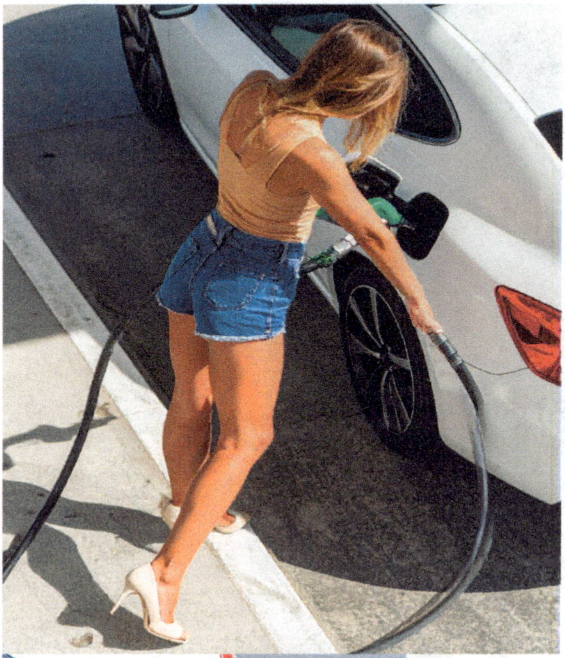

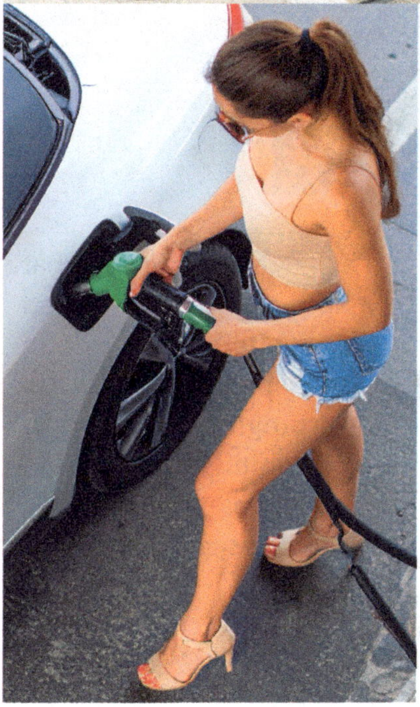

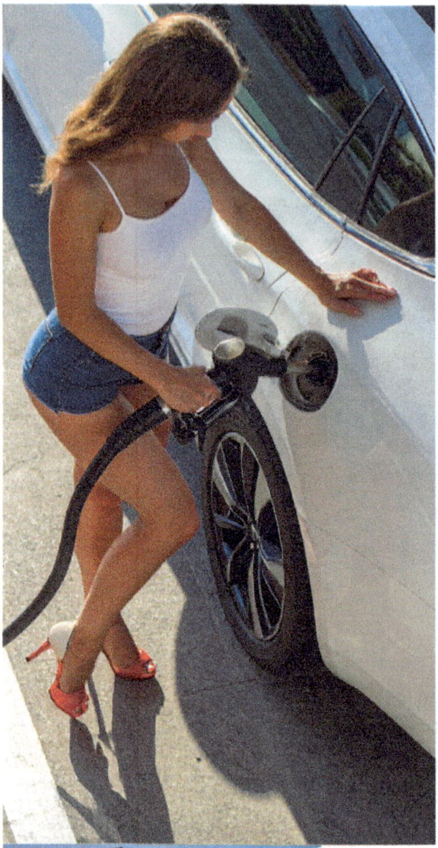

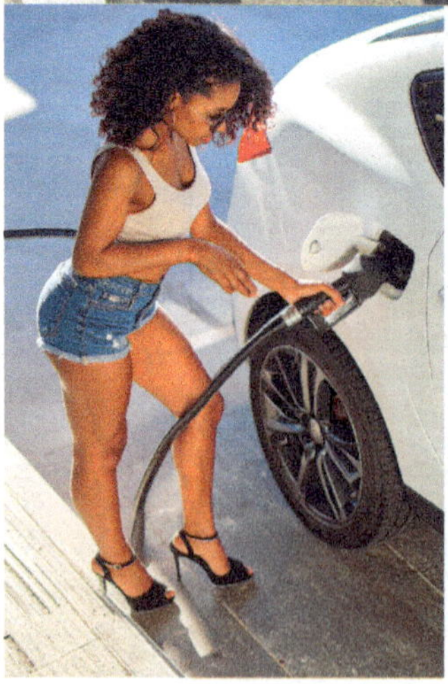

Not sure what made this 'white on white' color palette happen. Does the program take composition and complementary colors into account or not?

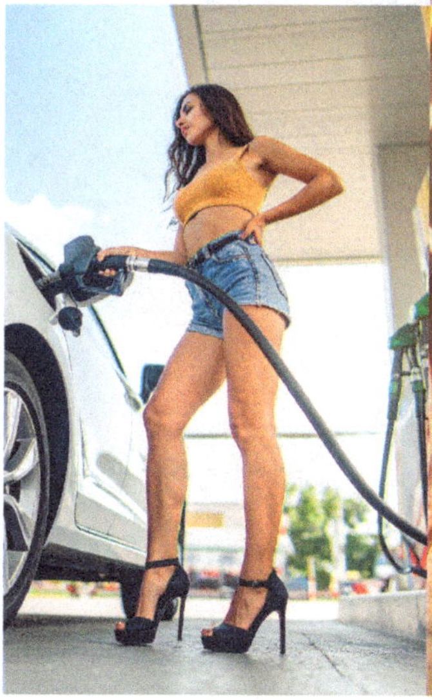

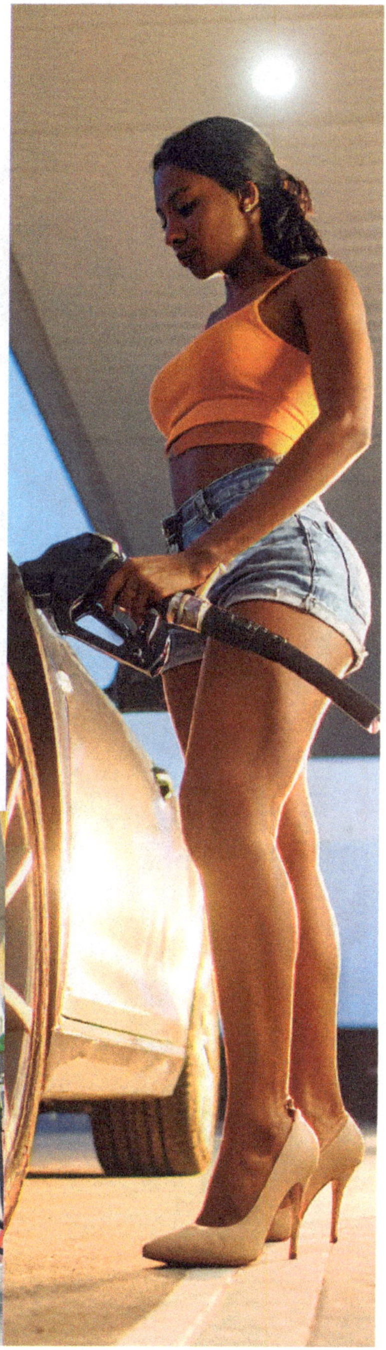

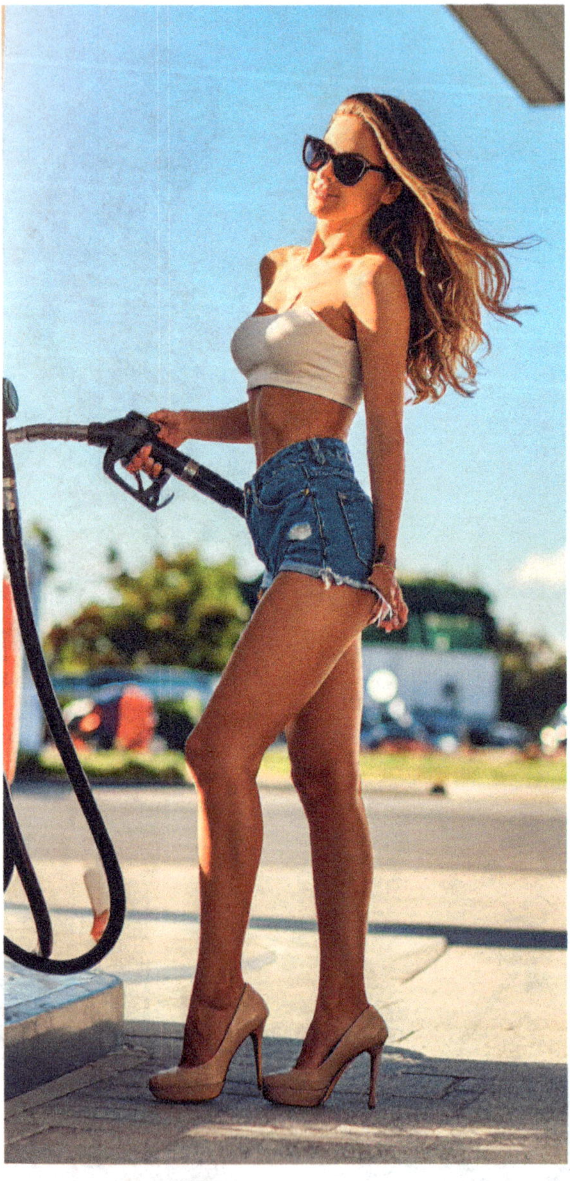

shot from above, a full body shot of sexy woman in tube top and skinny jean shorts and high heels pumping gas at gas station into fancy sports car

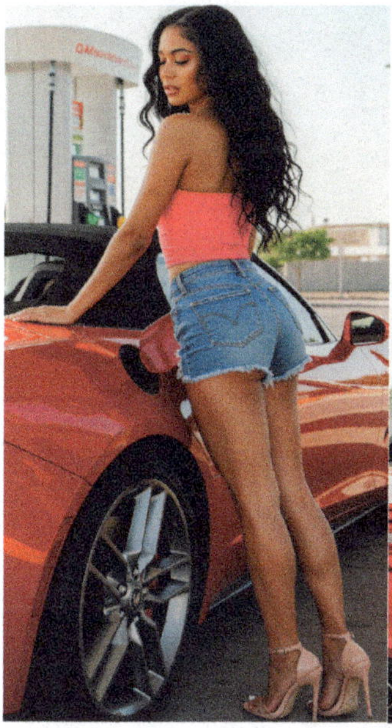

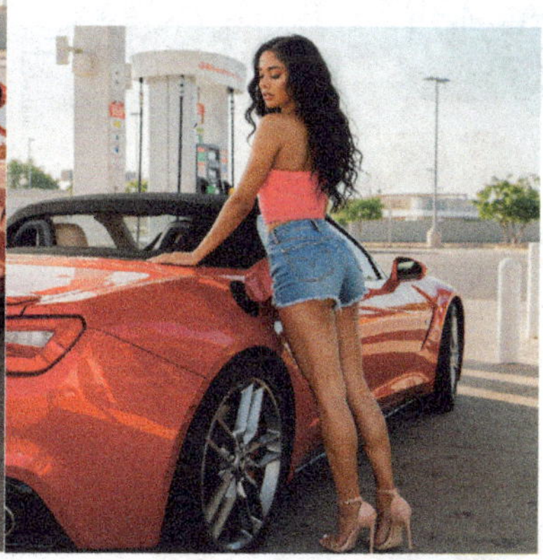

shot from below, a full body shot of sexy woman in tube top and skinny jean shorts and high heels pumping gas at gas station into fancy sports car

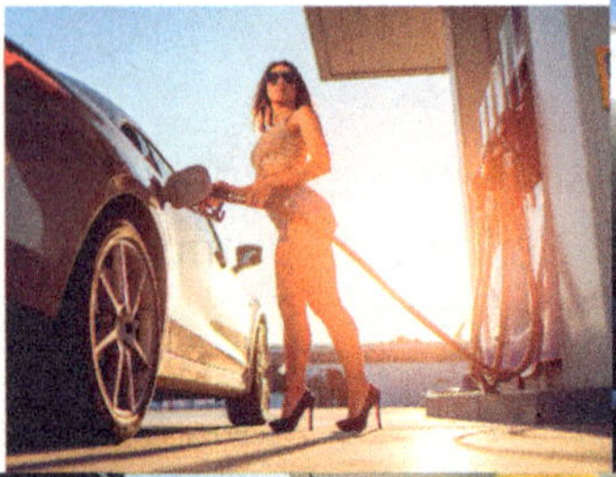

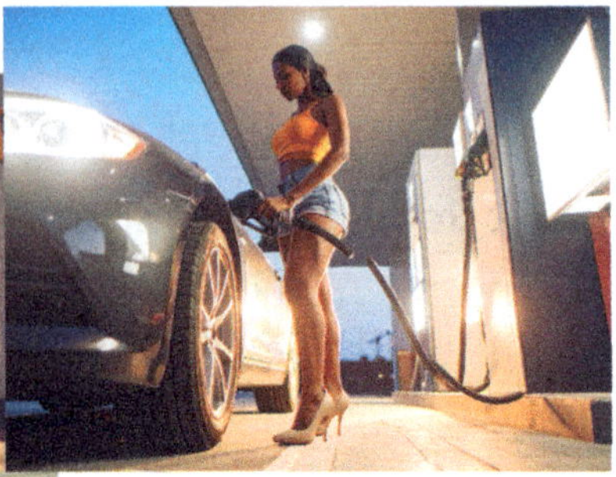

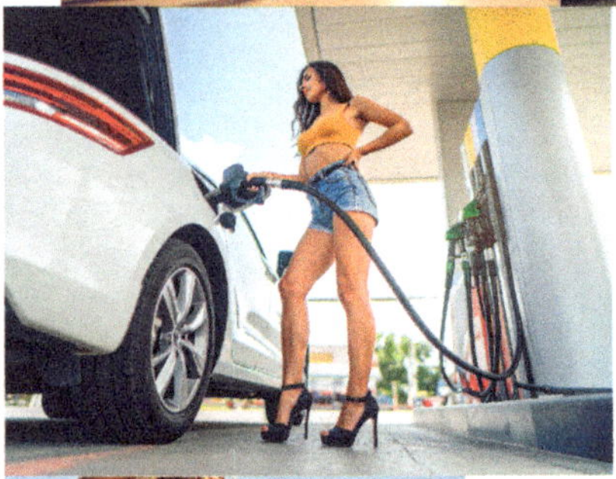

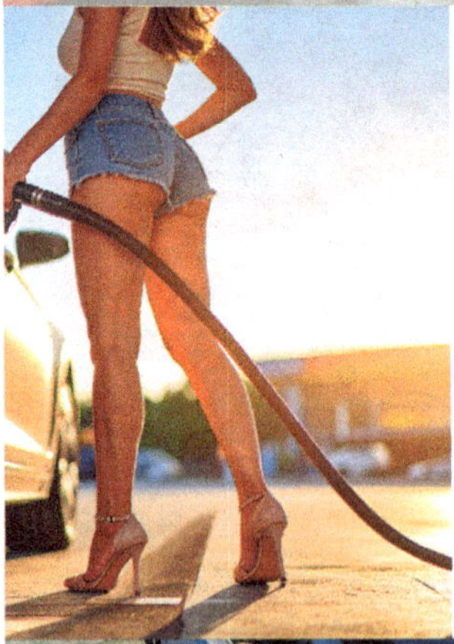

Why the bad lighting all of a sudden? And the muddled focus?

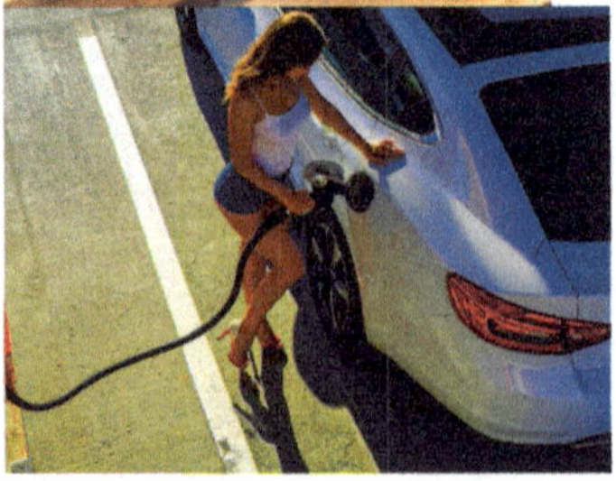

shot from above, a full body shot of sexy woman in tube top and skinny jeans and high heels pumping gas at gas station

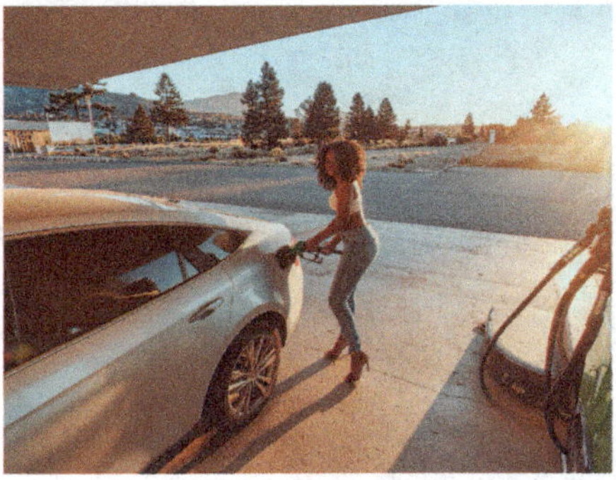

shot from above, a sexy woman in tube top and skinny jeans and high heels pumping gas at gas station

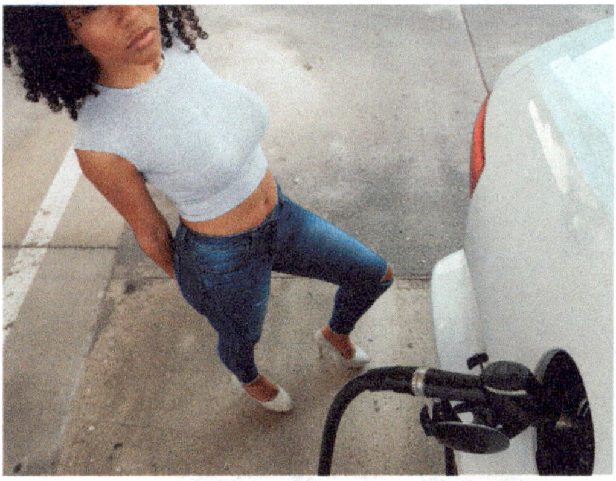

shot from above, a sexy woman in tube top and skinny jeans and high heels pumping gas at gas station

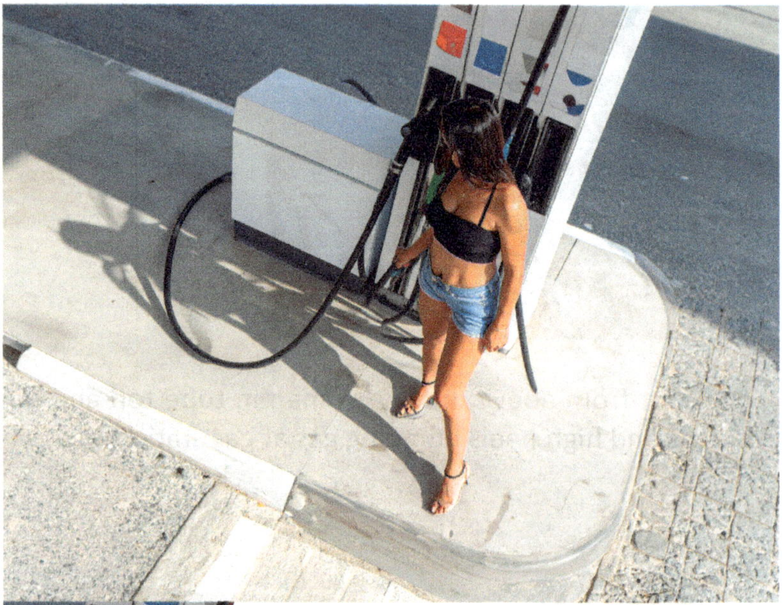

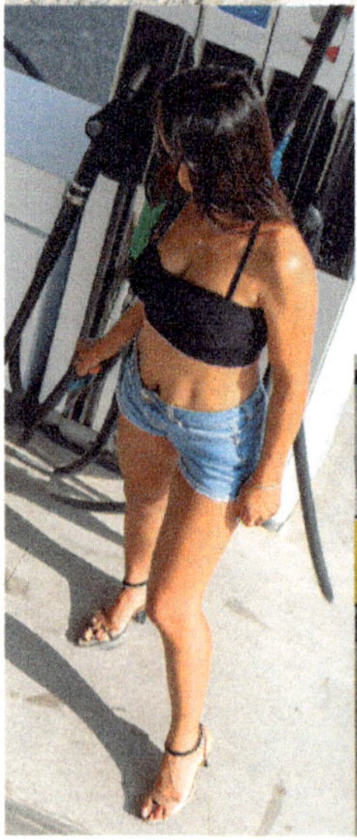

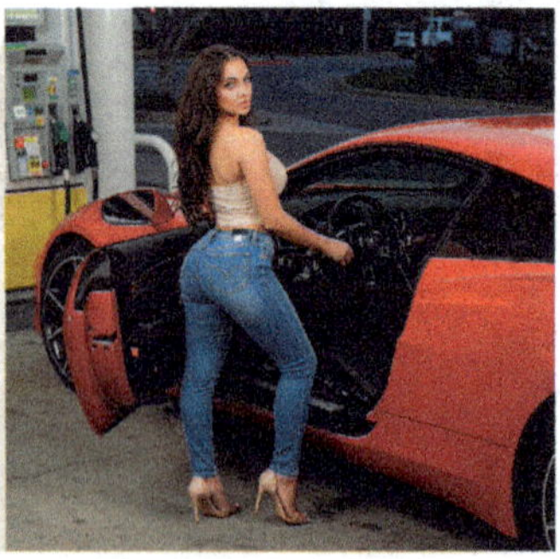

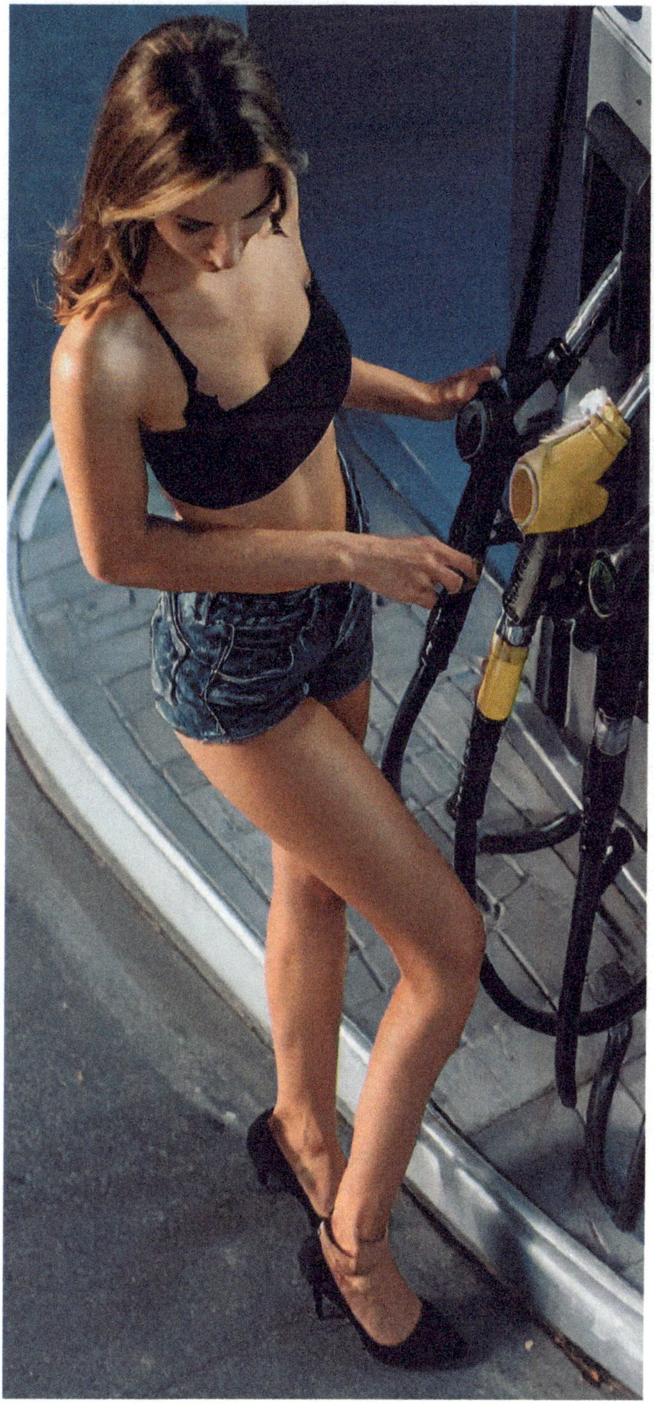

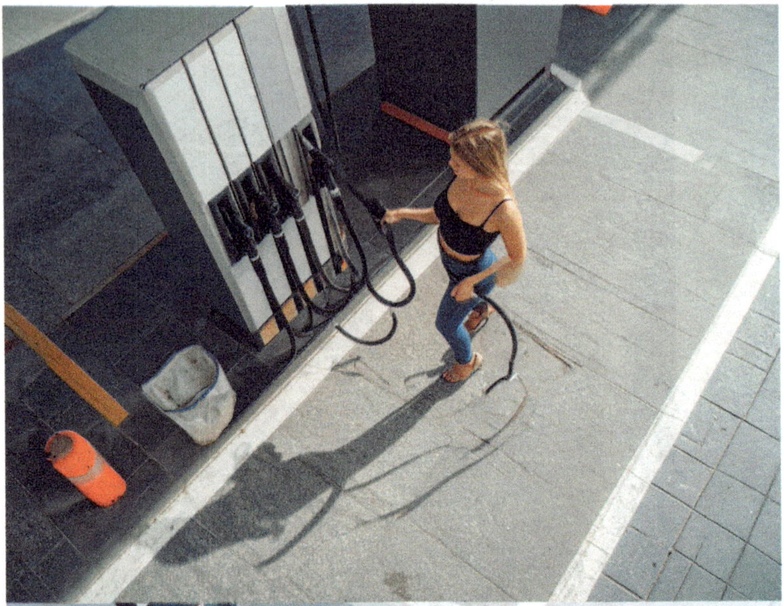
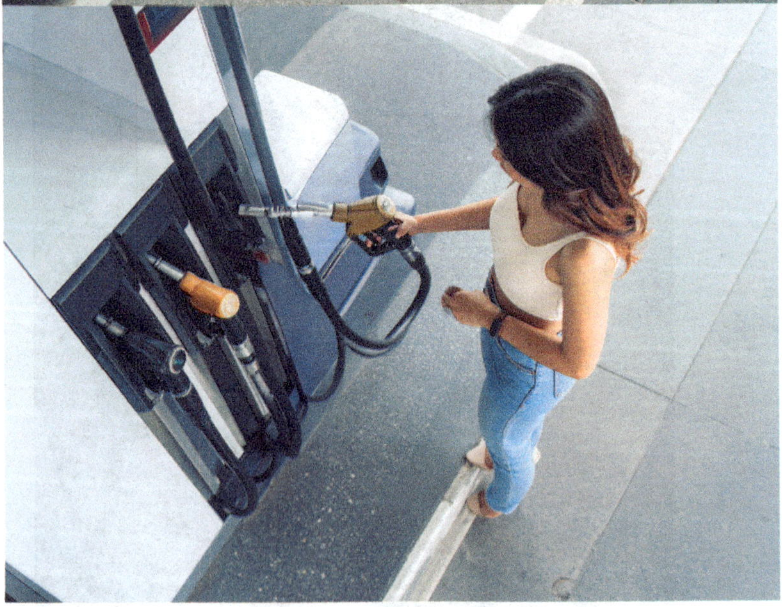

shot from above, a full body shot of sexy woman in tube top and skinny jeans and high heels pumping gas at gas station with beautiful women sitting in fancy sports car

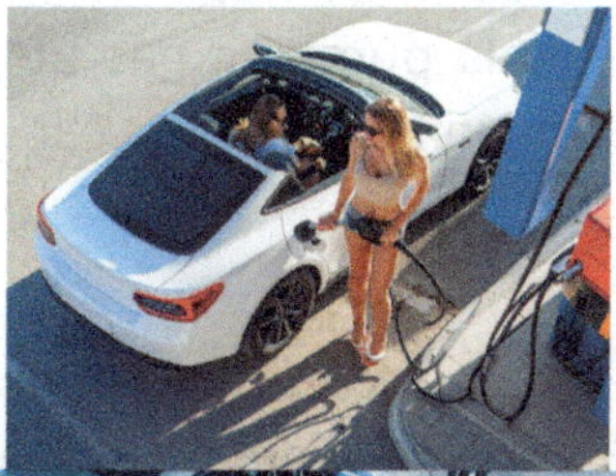

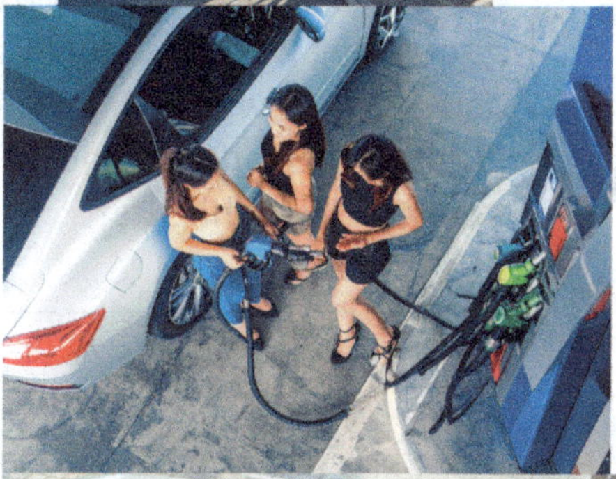

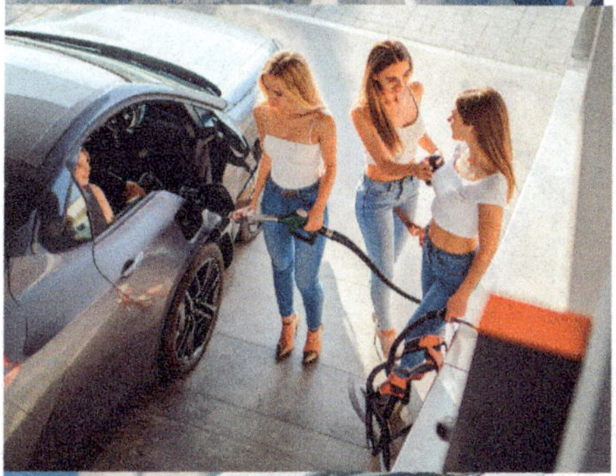

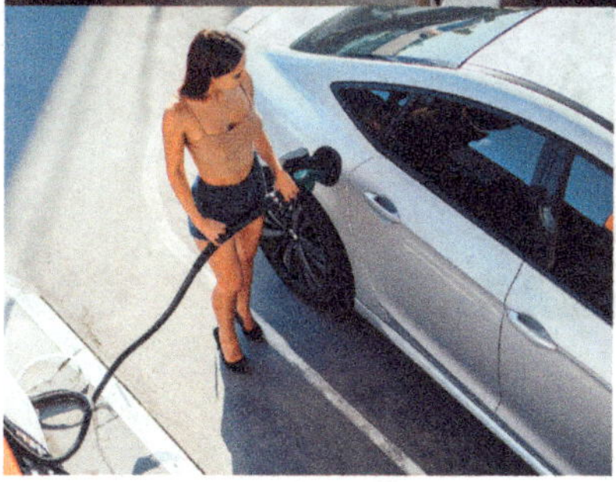

shot from above, a full body shot of sexy woman in tube top and skinny jeans and high heels pumping gas at gas station with beautiful women sitting in fancy sports car

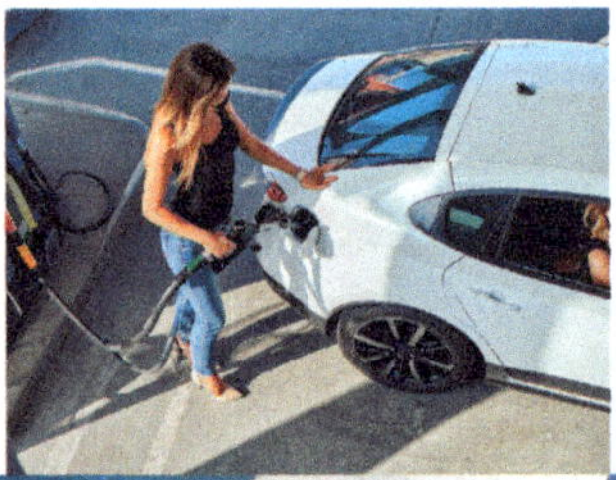

The AI seems to have trouble generating women actually in the car. Also, what iss with the confusing shadows and lack of contrasting highlights?

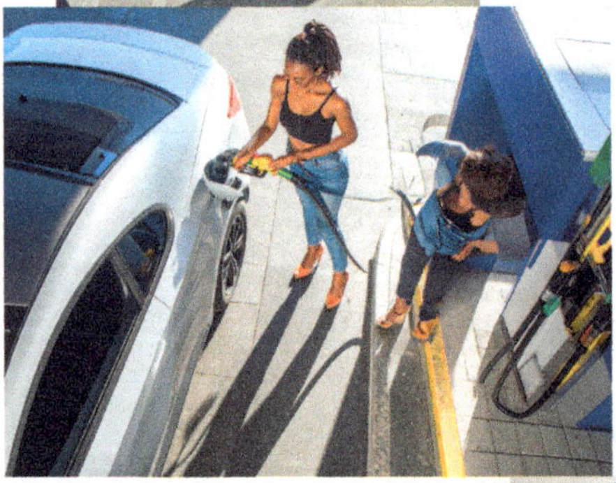

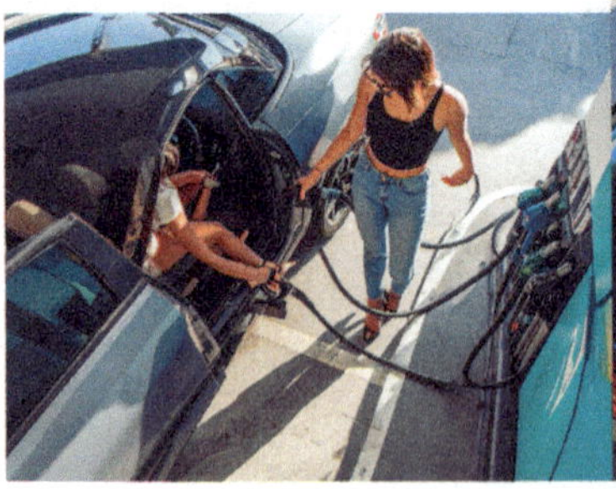

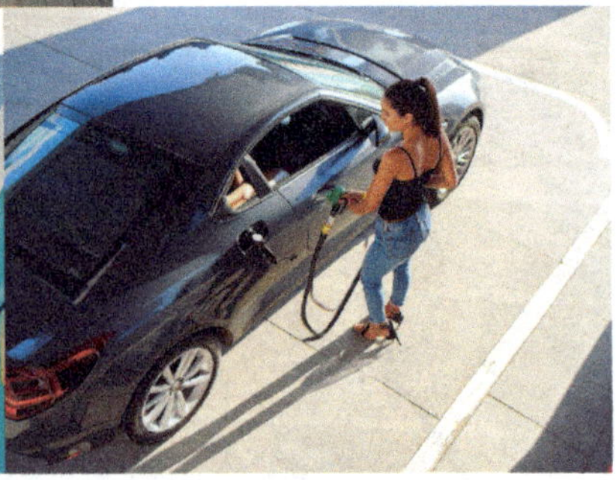

shot from above, a full body shot of sexy woman in tube top and skinny jeans and high heels pumping gas at gas station with beautiful women sitting in fancy convertible sports car

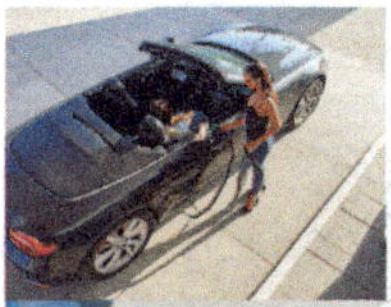

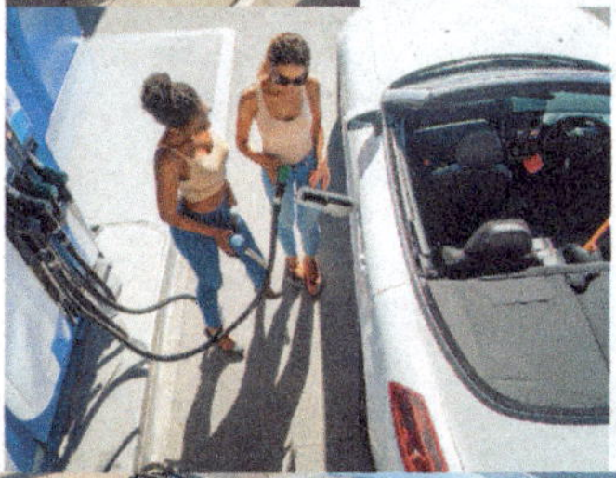

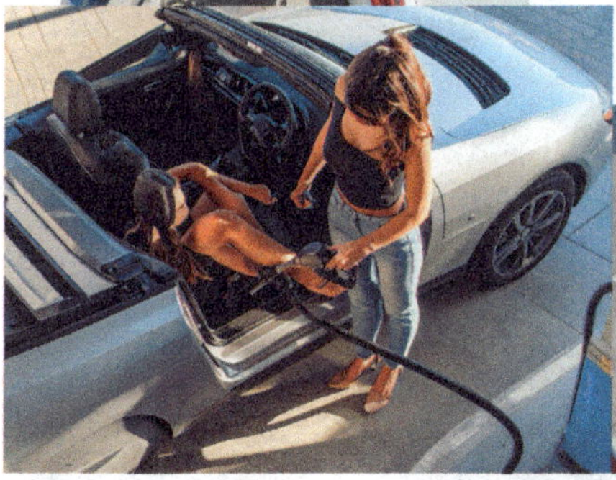

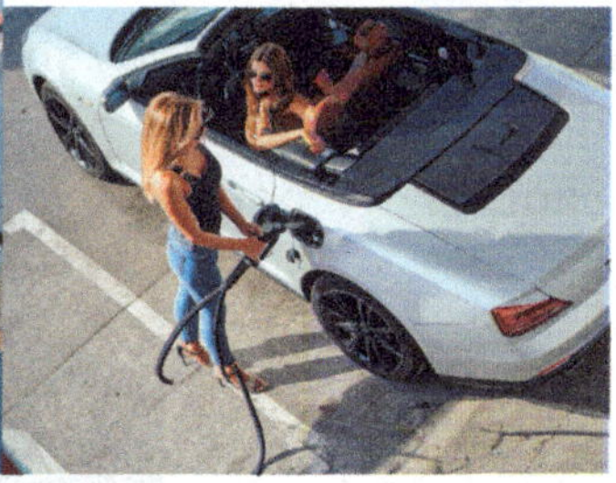

shot from above, a full body shot of sexy woman in tube top and skinny jeans and high heels pumping gas at gas station with beautiful women sitting in fancy convertible sports car

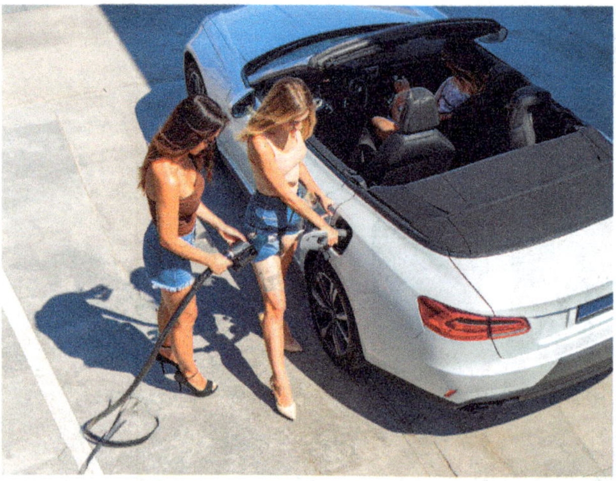

shot from above, a full body shot of sexy woman in tube top and skinny jeans and high heels pumping gas at gas station with beautiful women sitting in fancy convertible sports car

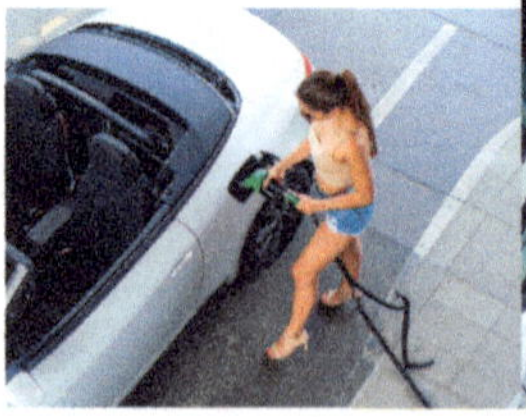

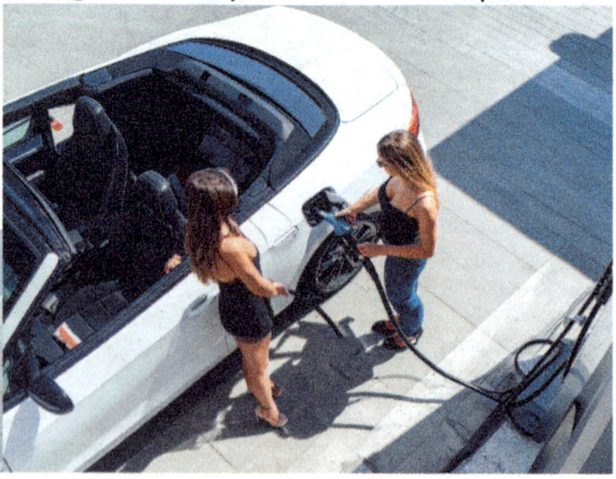

shot from above, a full body shot of sexy woman in tube top and skinny jeans and high heels pumping gas at gas station with beautiful women sitting in fancy convertible sports car

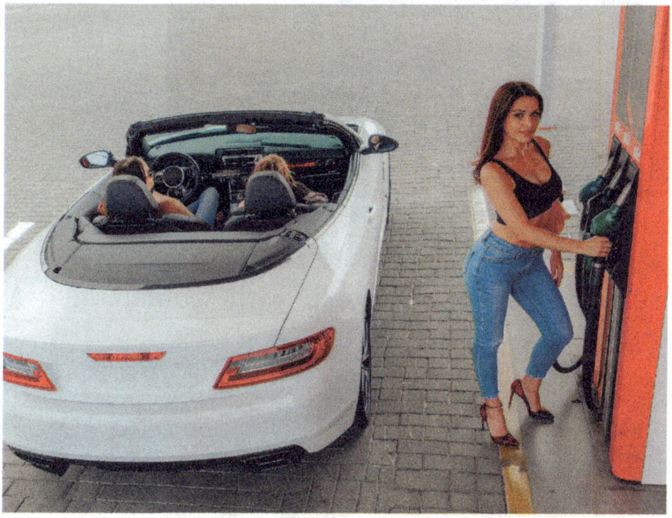

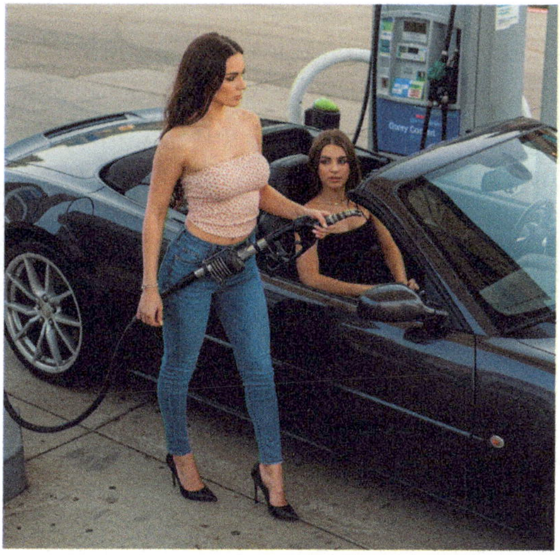

sexy woman button down shirt and miniskirt pumping gas at gas station into a convertible with women in it

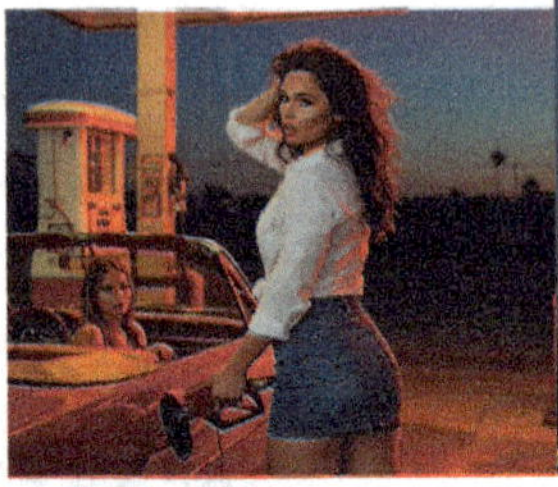
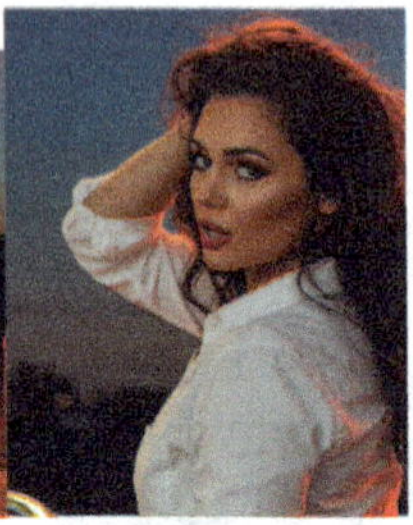

shot from the front, sexy woman button down shirt and miniskirt pumping gas at gas station into a convertible with women in it

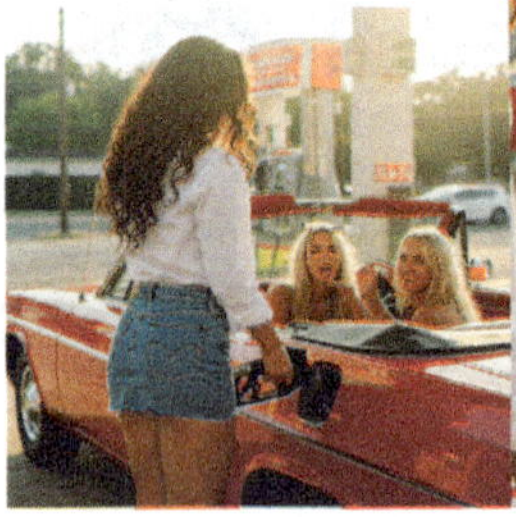
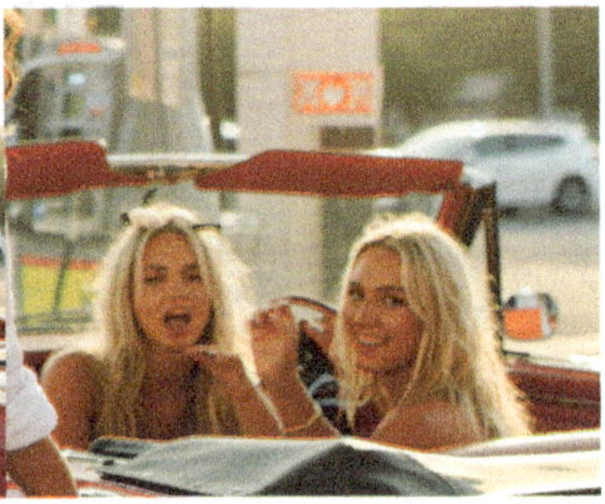

shot from the front, sexy woman in bathing suit pumping gas at gas station into a convertible with women in it

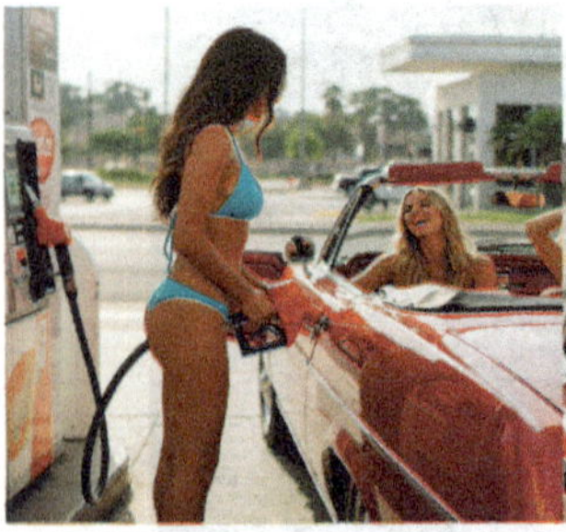
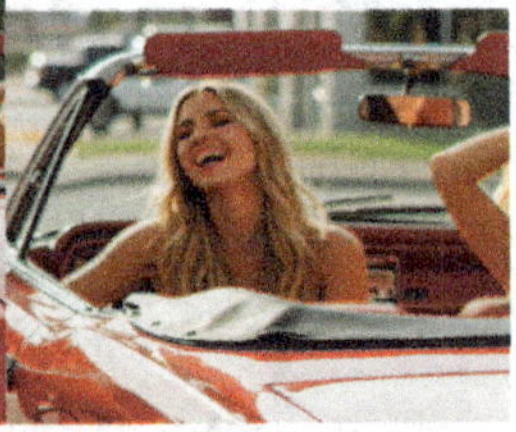

shot from the front, sexy woman in bathing suit pumping gas at gas station into a convertible with women in it

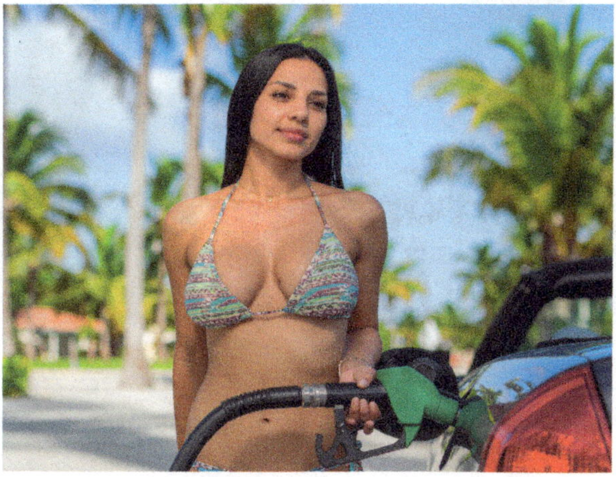

woman in string bikini pumping gas at gas station into sports car

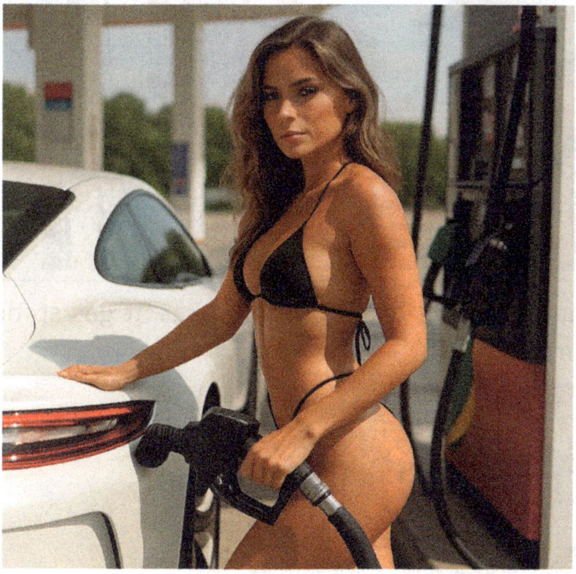

woman in string bikini pumping gas at gas station into sports car

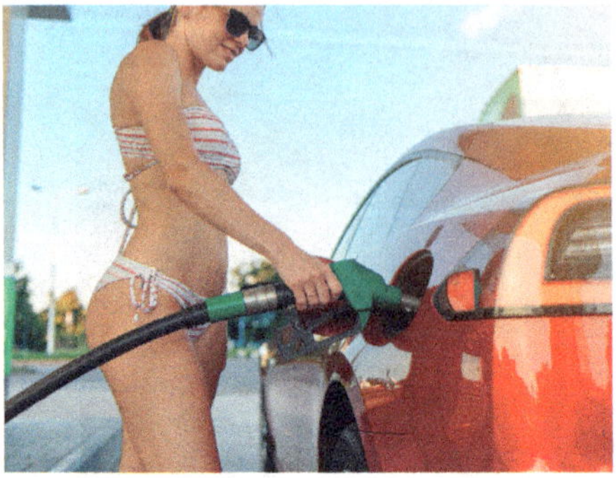

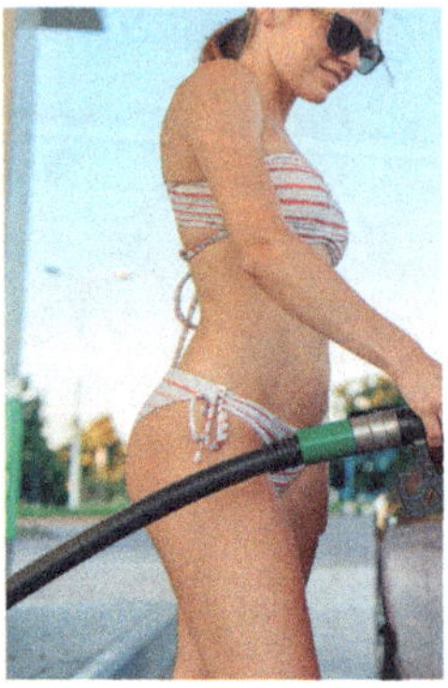

woman in string bikini pumping gas at gas station into sports car

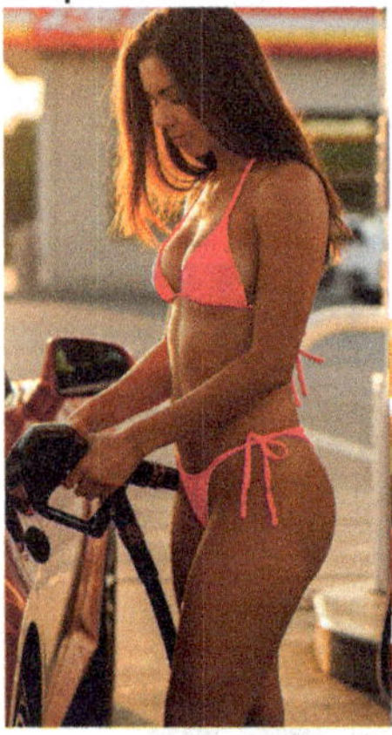

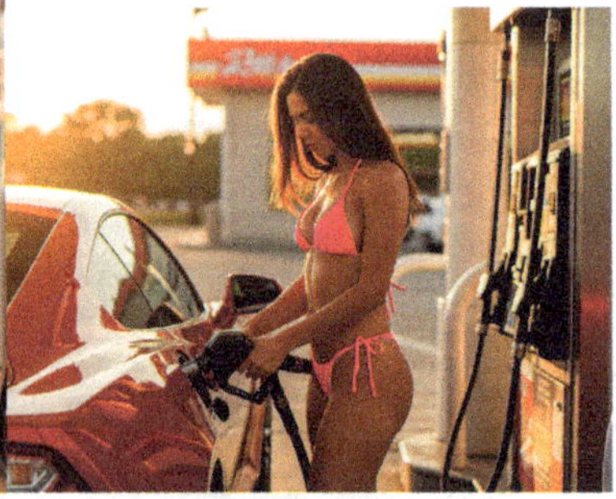

woman in string bikini pumping gas at gas station into sports car

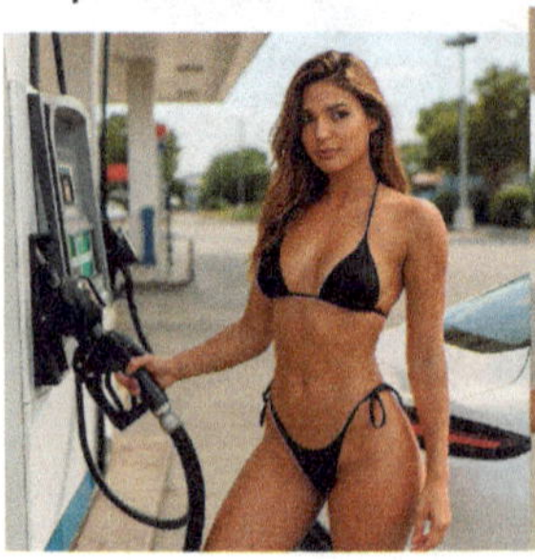

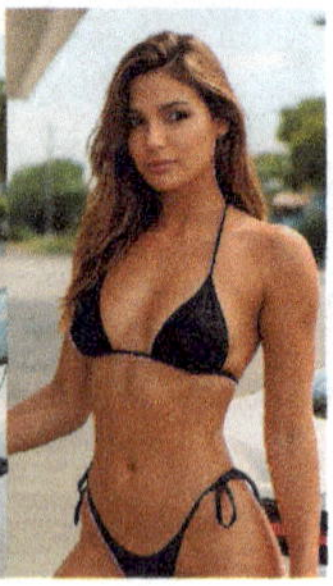

woman in string bikini pumping gas at gas station into sports car and posing for camera

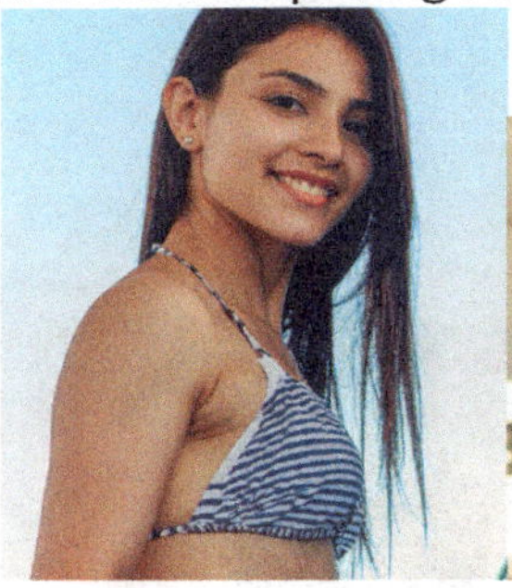

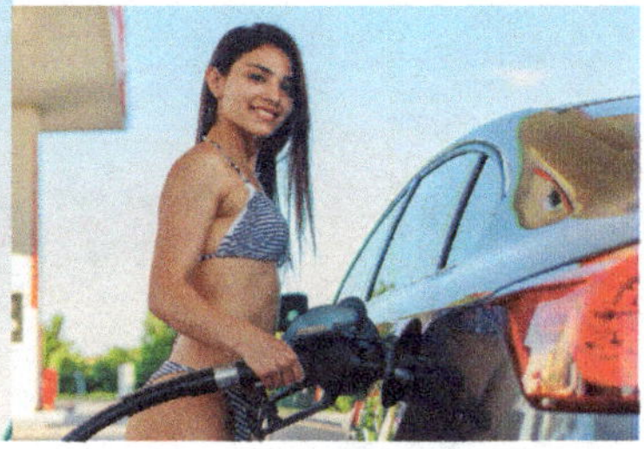

tall, buxom, sexy woman pumping gas at gas station

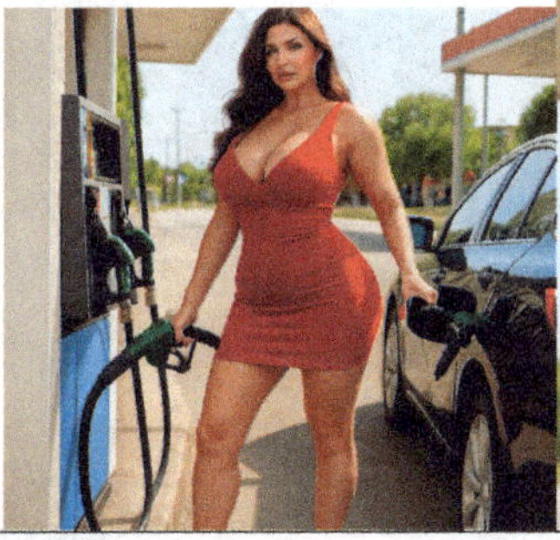

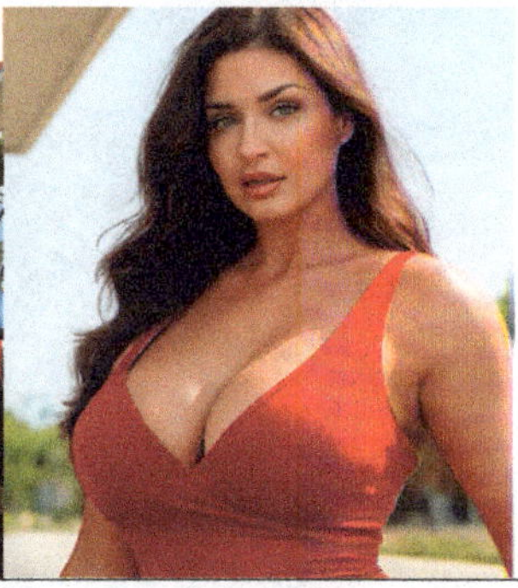

tall, skinny, buxom, sexy woman pumping gas at gas station

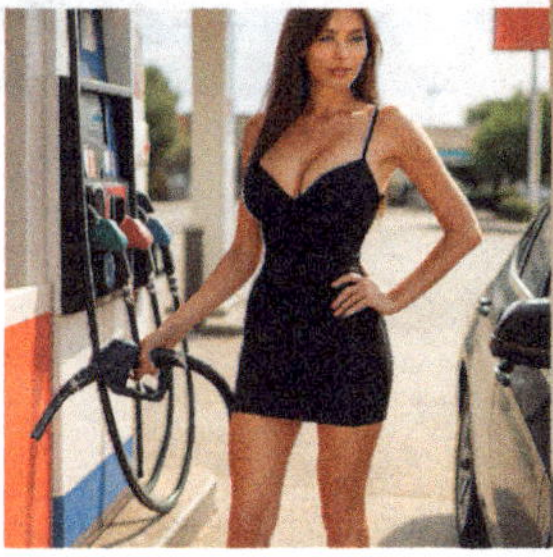

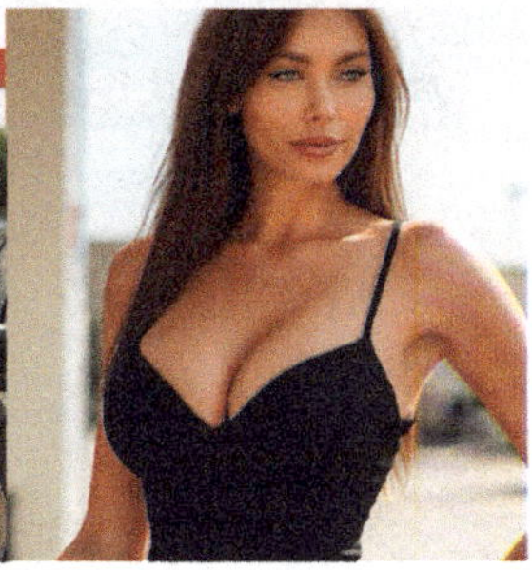

tall, skinny, buxom, sexy woman pumping gas at gas station

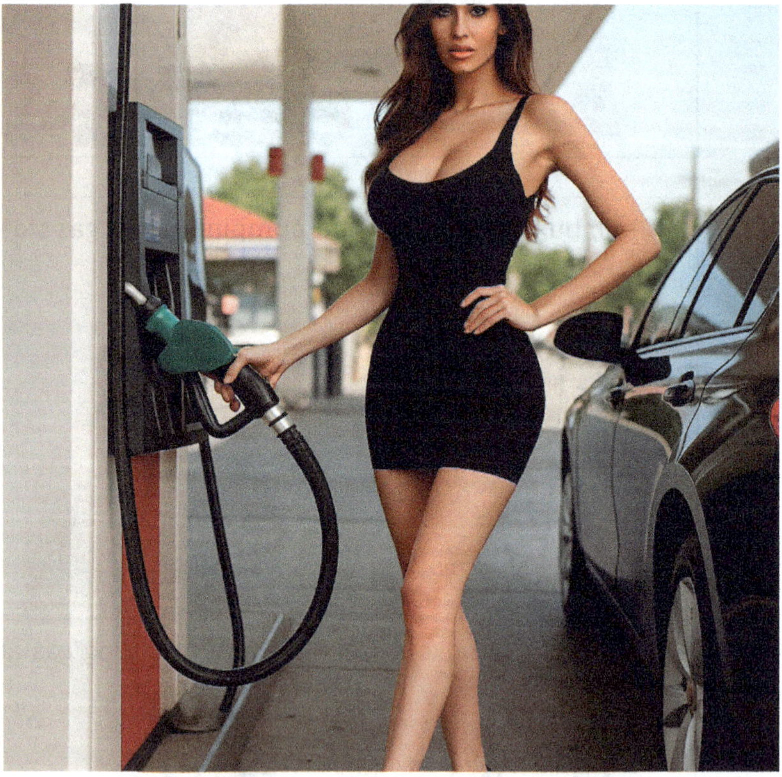

tall, skinny, buxom, sexy woman pumping gas at gas station

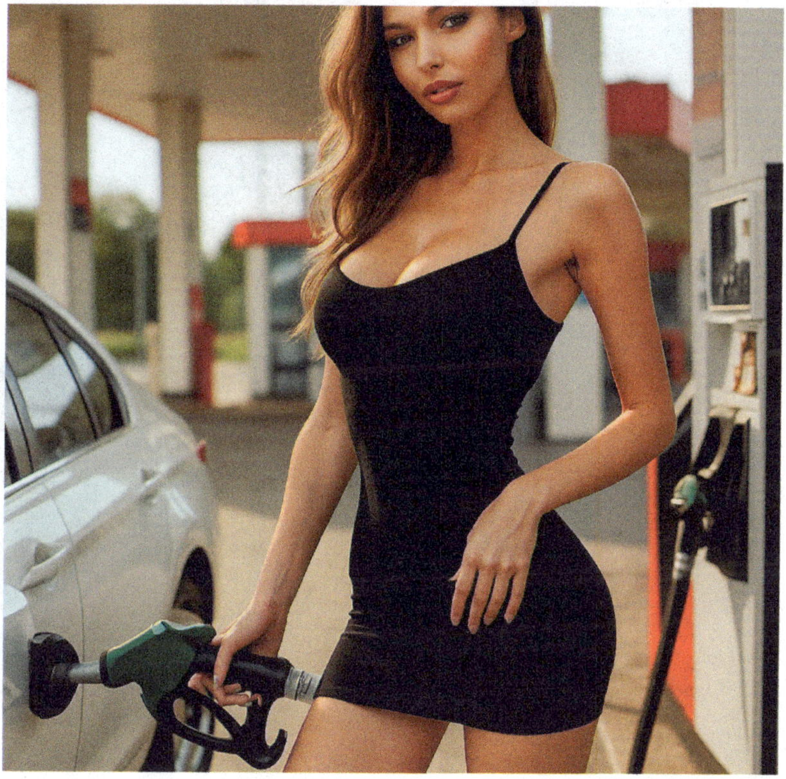

tall, skinny, buxom, sexy woman pumping gas at gas station

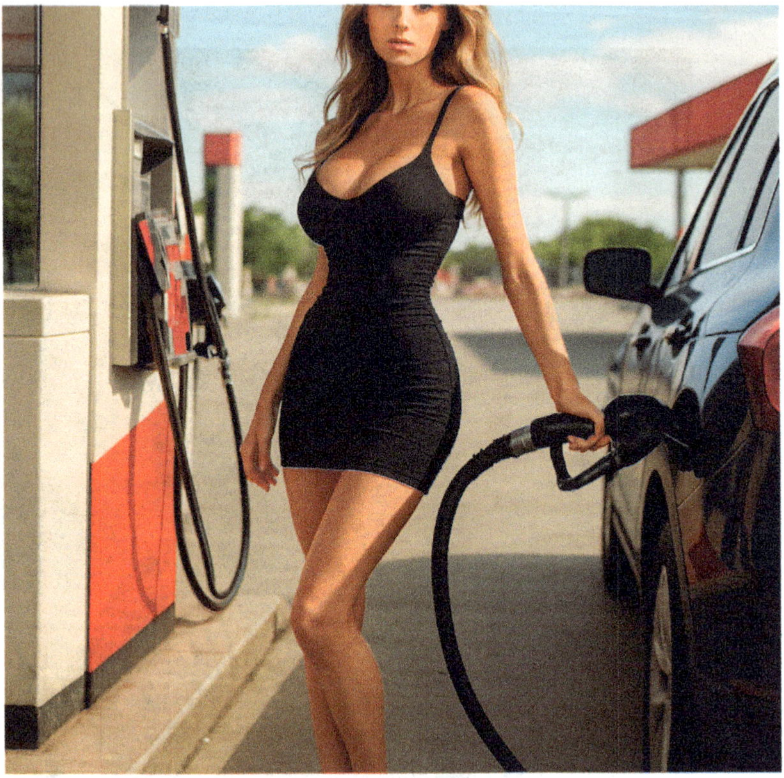

tall, skinny, buxom, sexy woman wearing a bathing suit, pumping gas at gas station

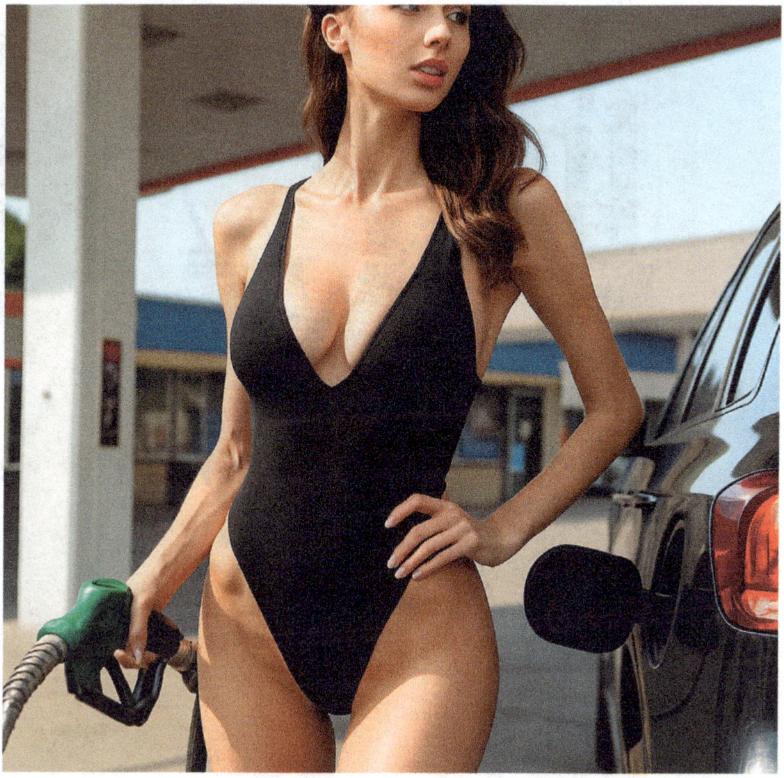

tall, skinny, buxom, sexy woman wearing a bathing suit, pumping gas at gas station

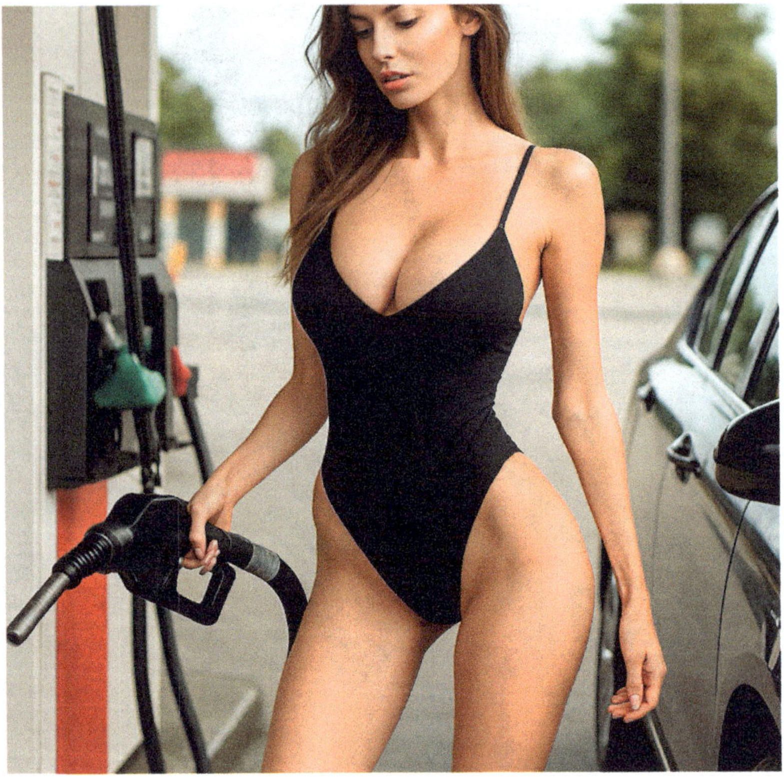

tall, skinny, buxom, sexy woman wearing a bathing suit, pumping gas at gas station

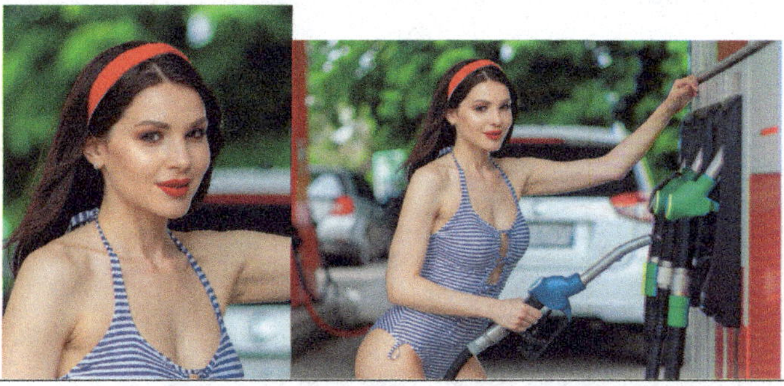

tall, skinny, buxom, sexy woman wearing a bathing suit, pumping gas at gas station

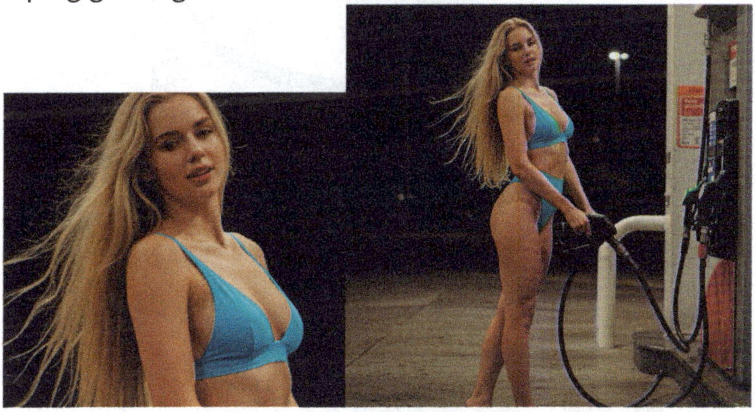

shot from above, tall, skinny, buxom, sexy woman wearing a bathing suit, pumping gas at gas station into fancy convertible car with beautiful women

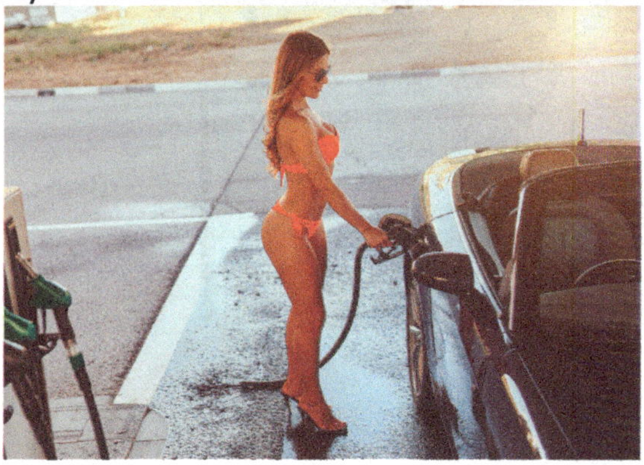

shot from above, skinny, buxom, woman wearing a bathing suit, pumping gas at gas station

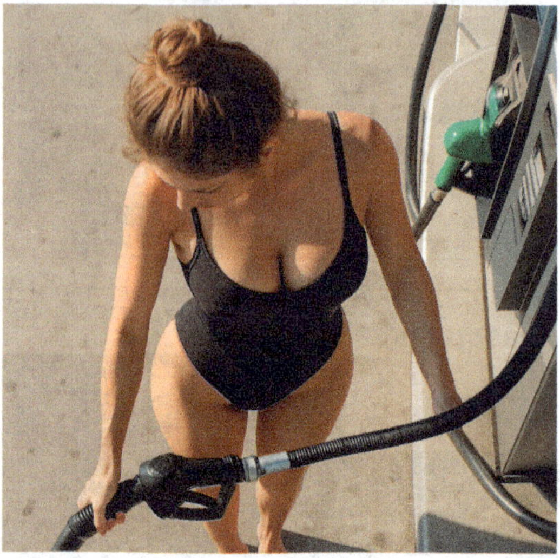

shot from above, skinny, buxom, woman wearing a bathing suit, pumping gas at gas station

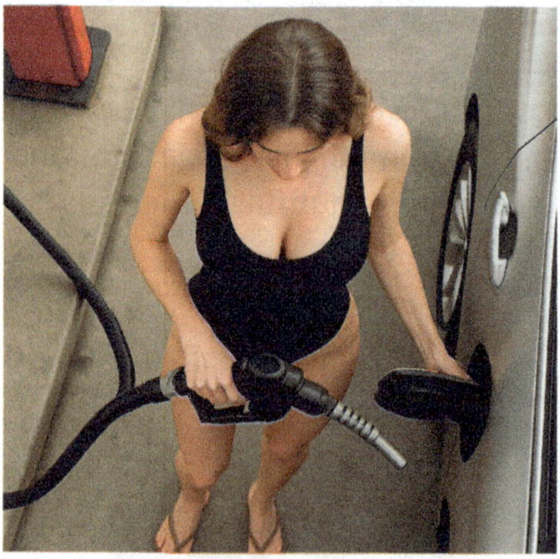

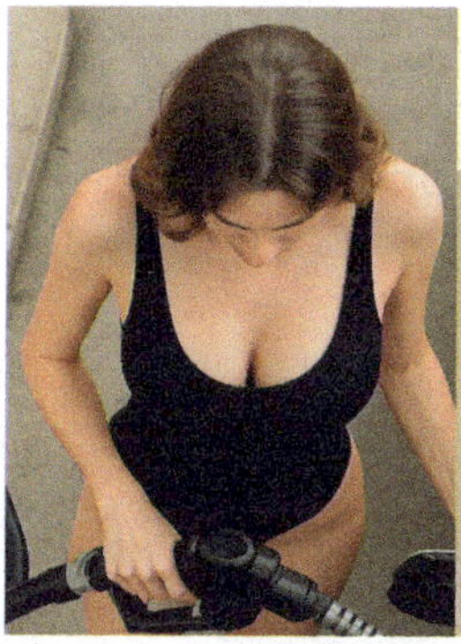
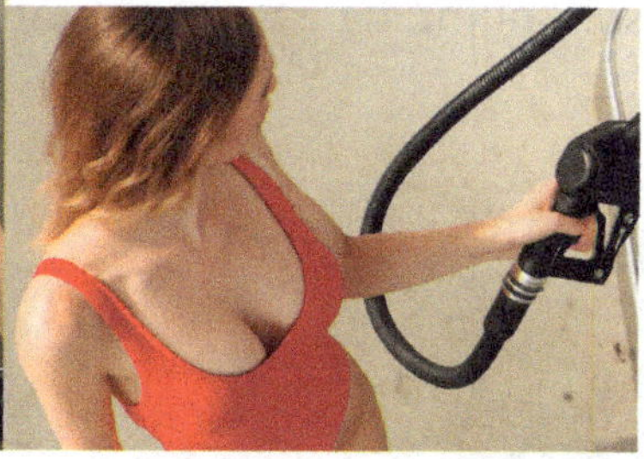
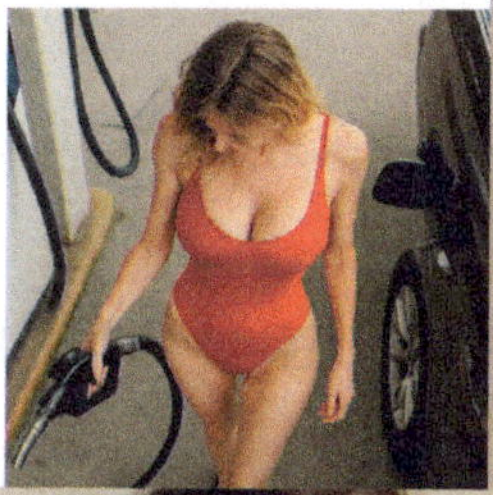
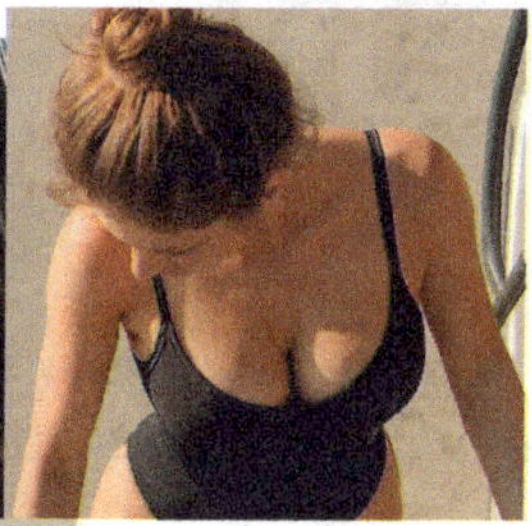
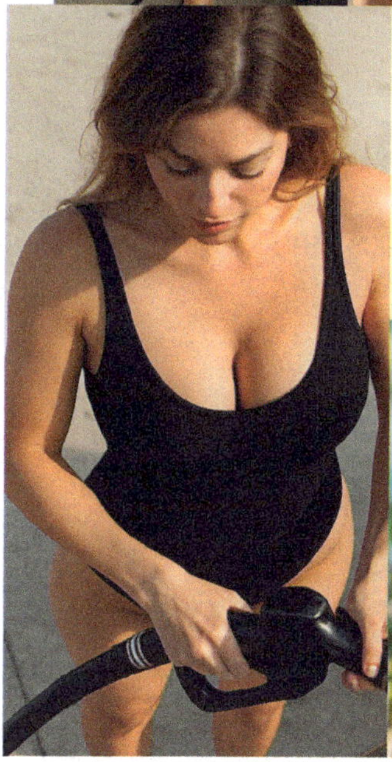
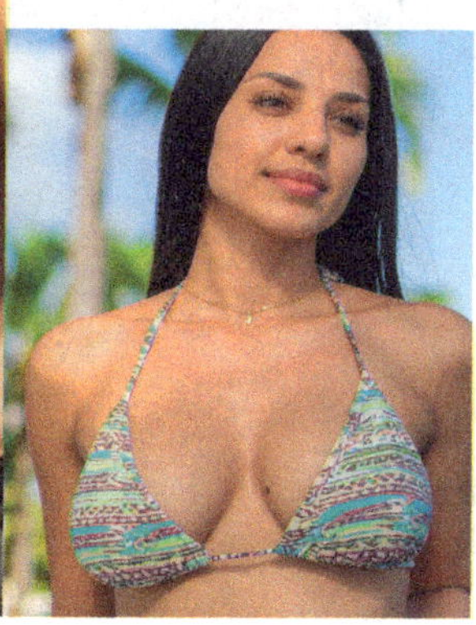

shot from above, skinny, buxom, woman wearing a bathing suit, pumping gas at gas station

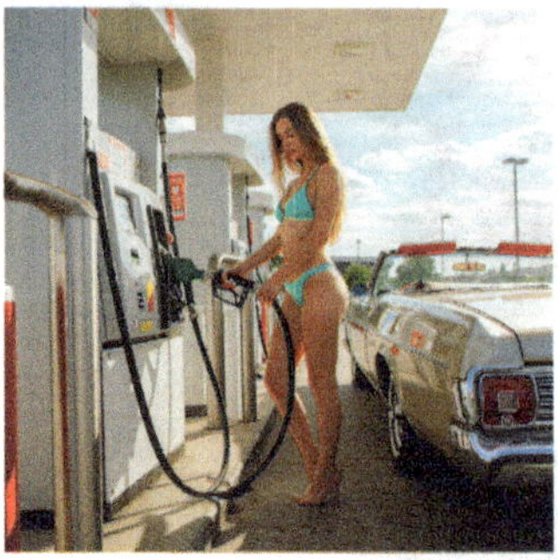

shot from above, skinny, buxom, woman wearing a bathing suit, pumping gas at gas station

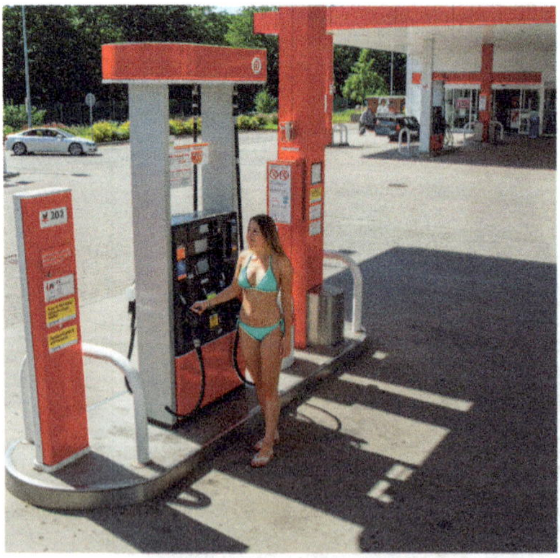

shot from above, skinny, buxom, woman wearing a bathing suit, pumping gas at gas station

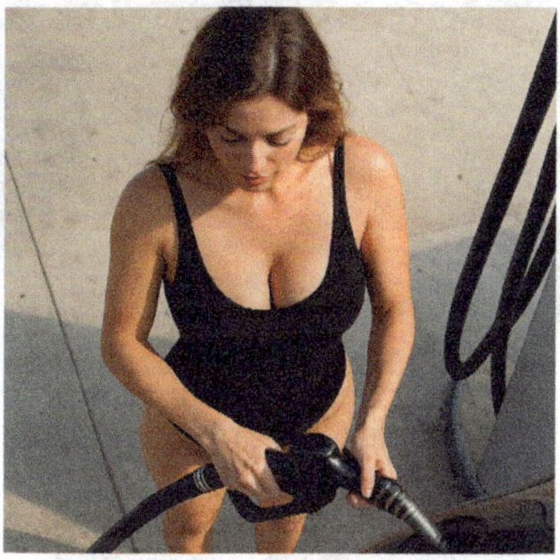

shot from below, skinny, buxom, woman wearing a bathing suit, pumping gas at gas station

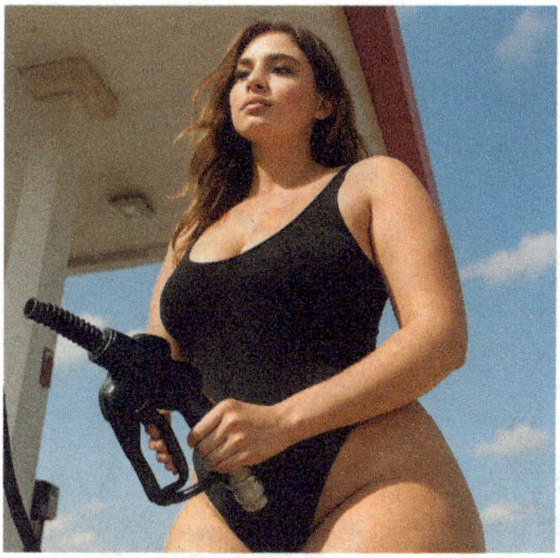

shot from below, skinny, buxom, woman wearing a bathing suit, pumping gas at gas station

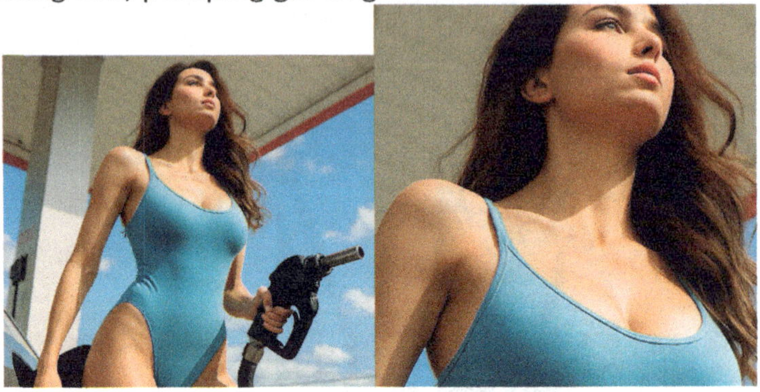

shot from below, skinny, buxom, woman wearing a bathing suit, pumping gas at gas station

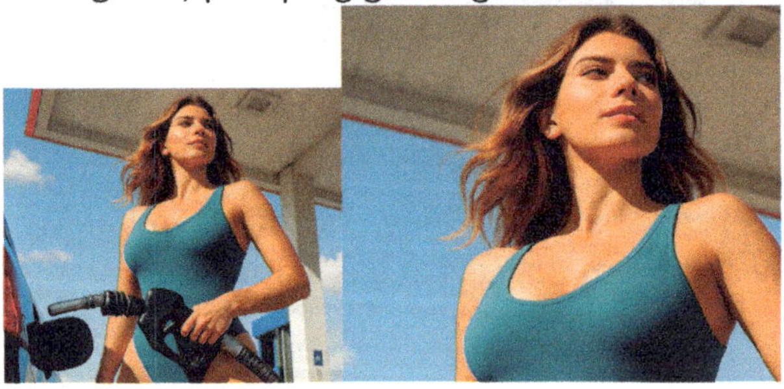

shot from below, skinny, buxom, woman wearing a bathing suit, pumping gas at gas station

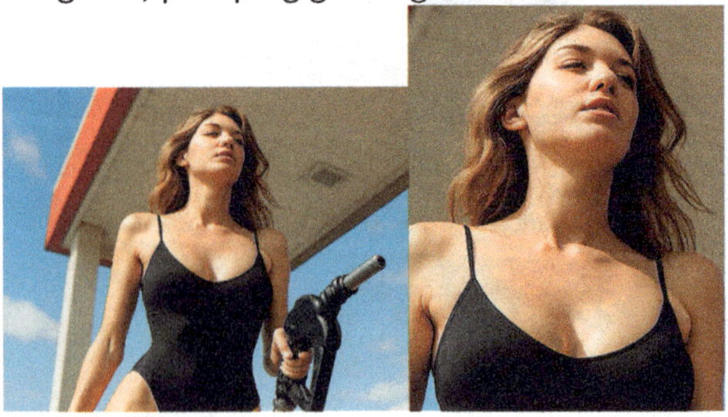

shot from below, skinny, buxom, woman wearing a bathing suit, pumping gas at gas station

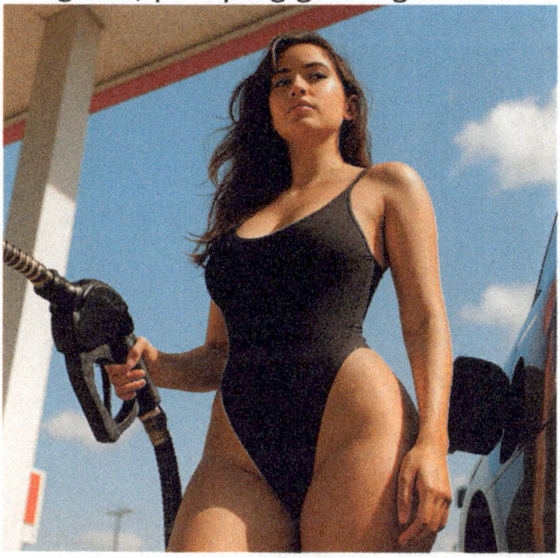

shot from below, skinny, buxom, woman wearing a string bikini bathing suit, pumping gas at gas station

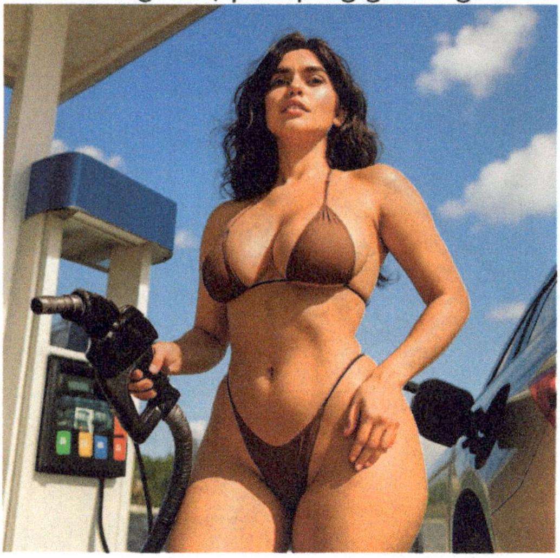

shot from below, skinny, buxom, woman wearing a bathing suit, pumping gas at gas station

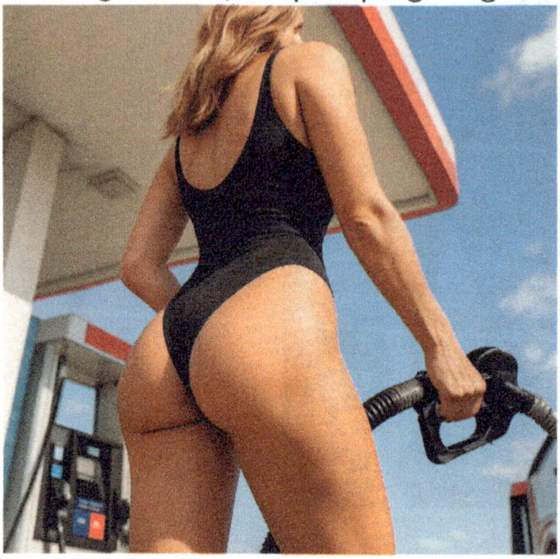

shot from above, skinny, buxom, woman wearing a bathing suit, pumping gas at gas station

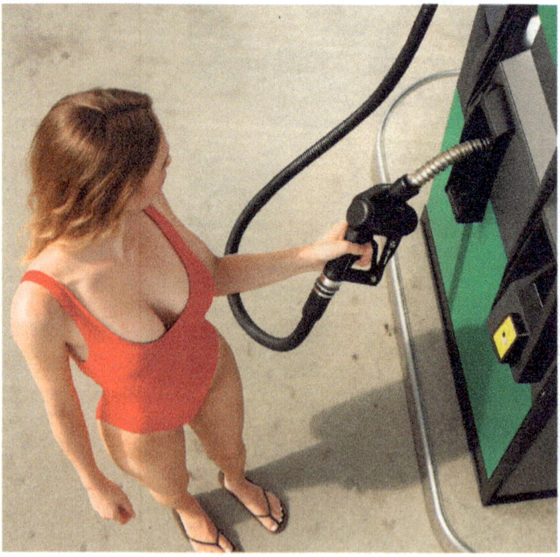

shot from below, skinny, buxom, woman wearing a bathing suit, pumping gas at gas station

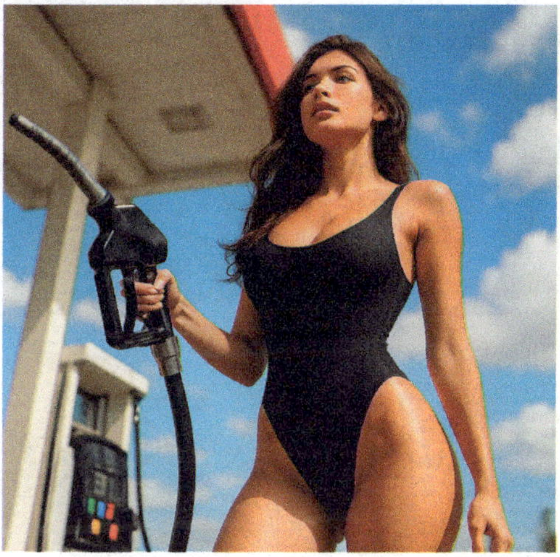

shot from above, skinny, buxom, woman wearing a bathing suit, pumping gas at gas station

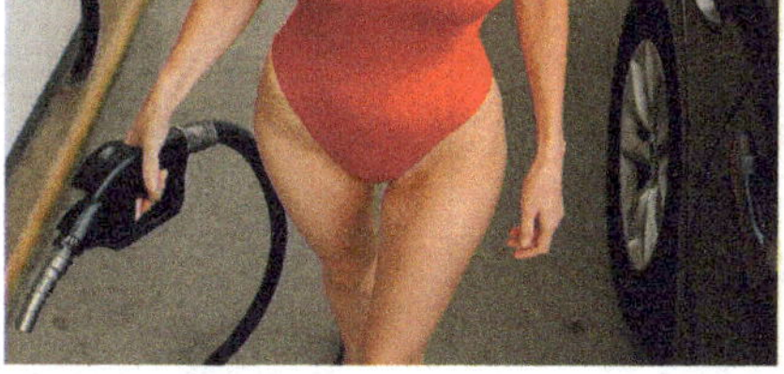

shot from above, skinny, buxom, woman wearing a bathing suit, pumping gas at gas station

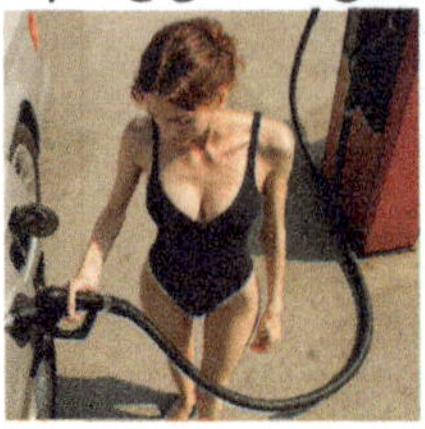

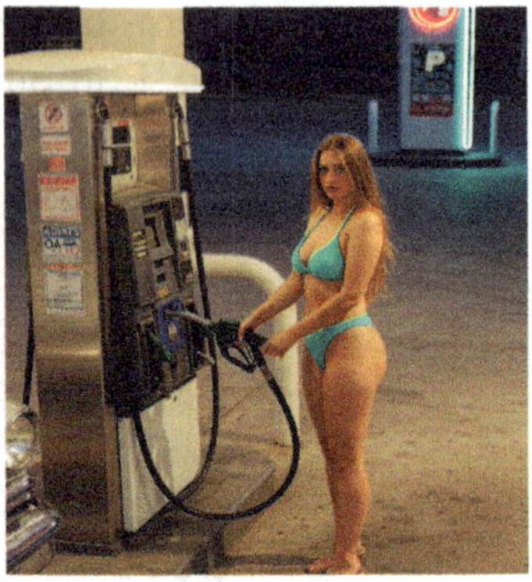

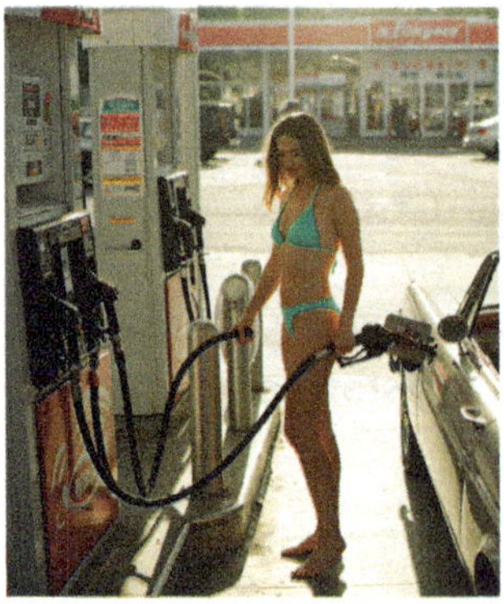

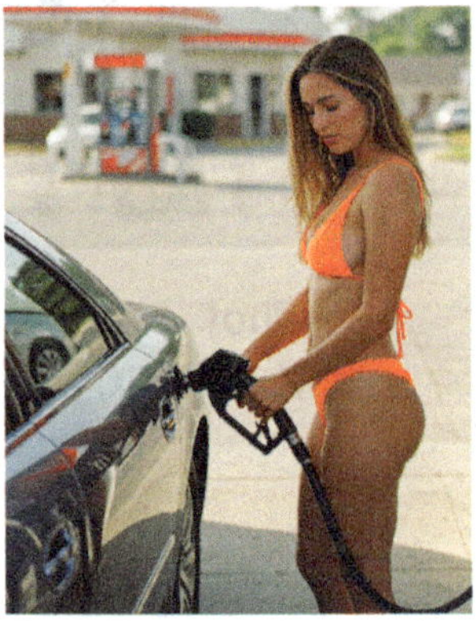

orange bathing suit,

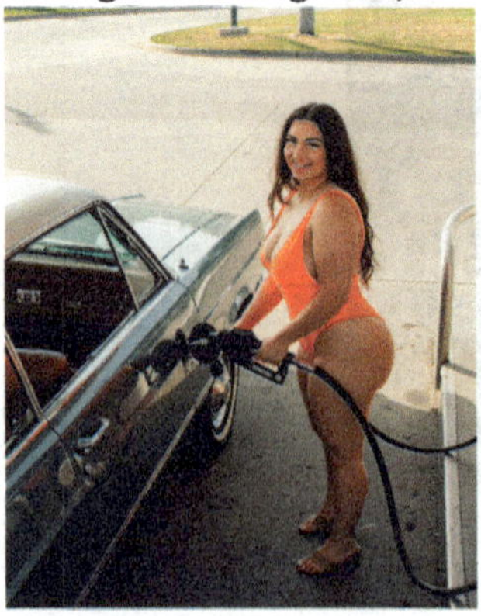

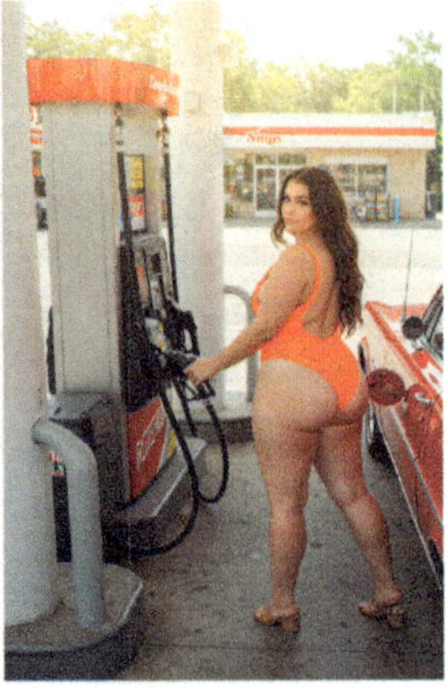

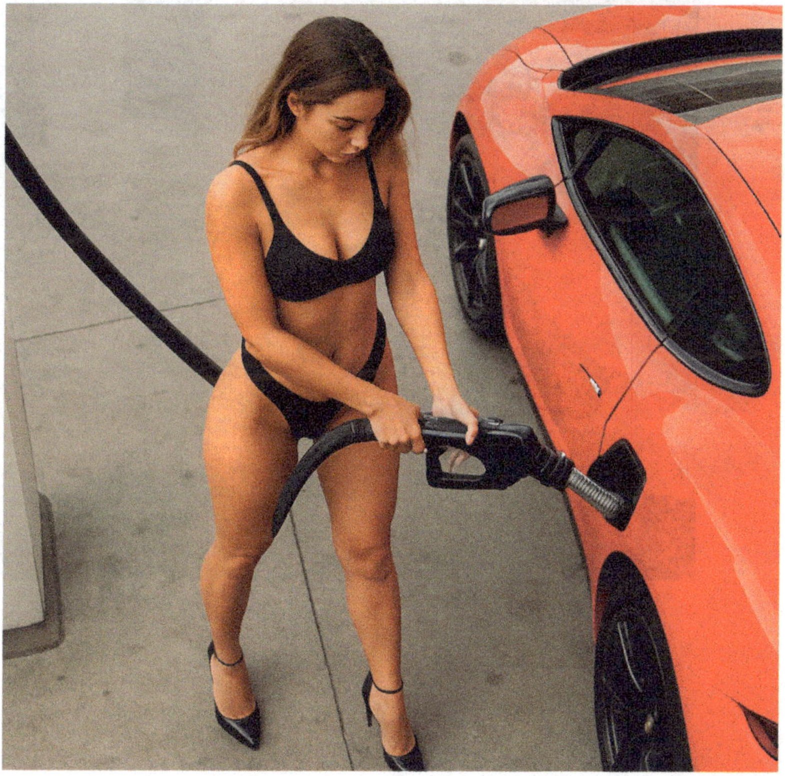

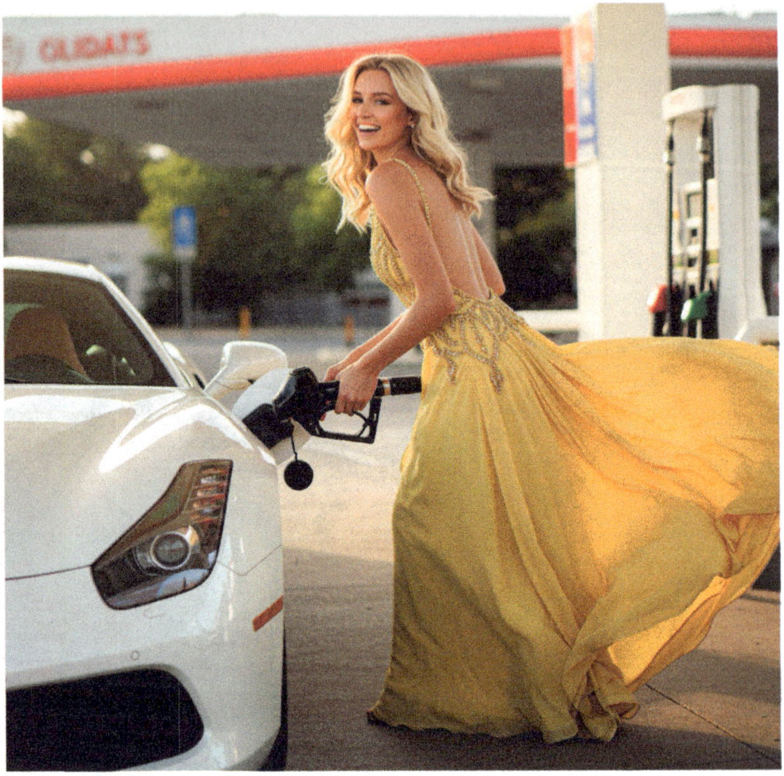

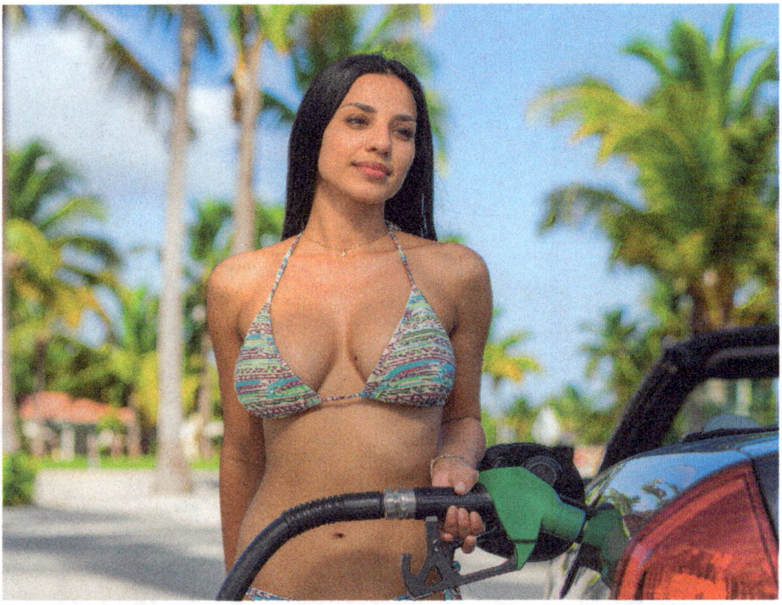

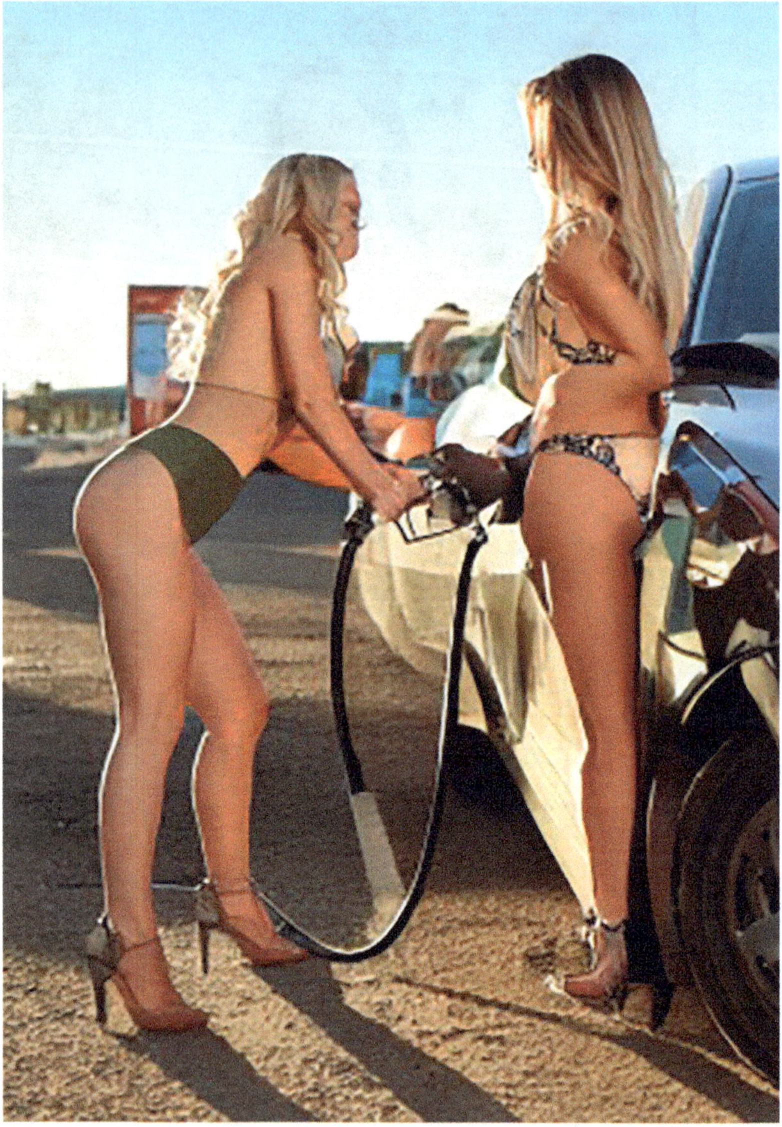

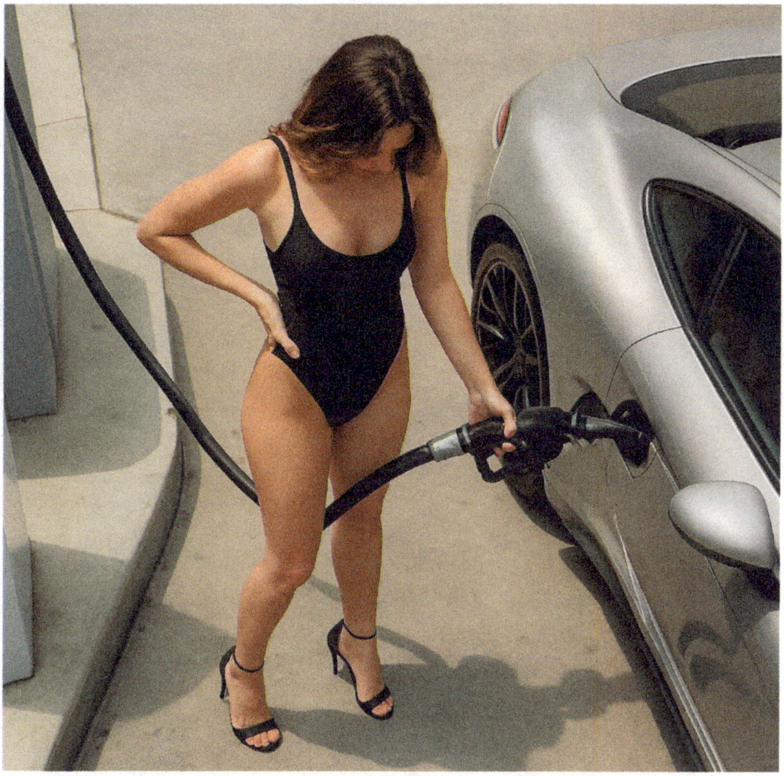

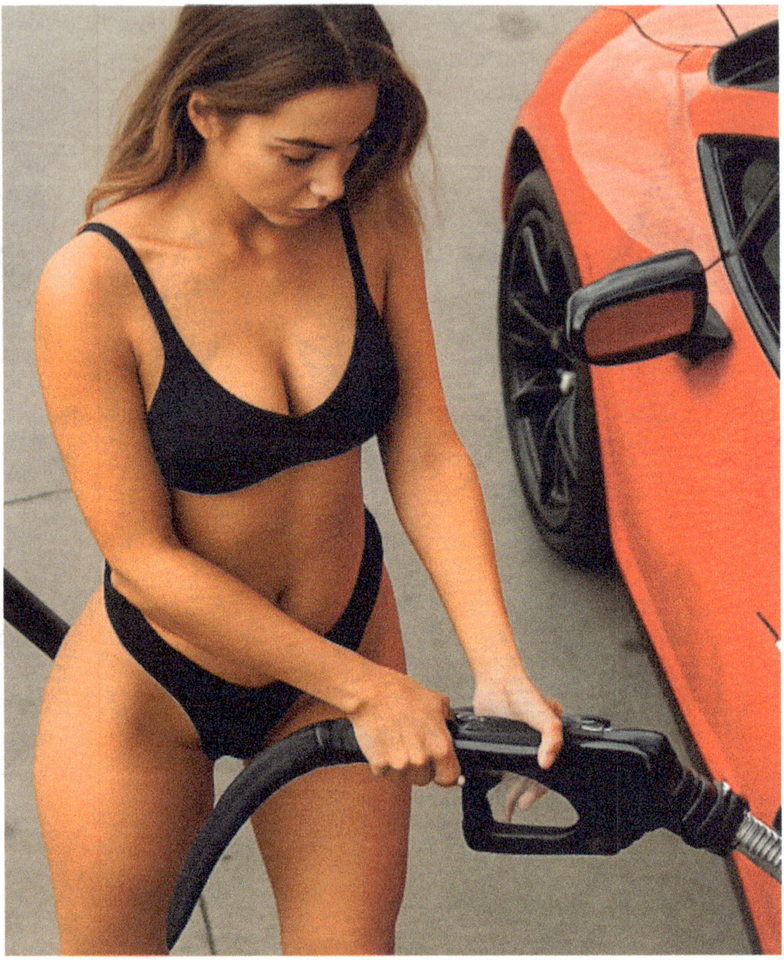

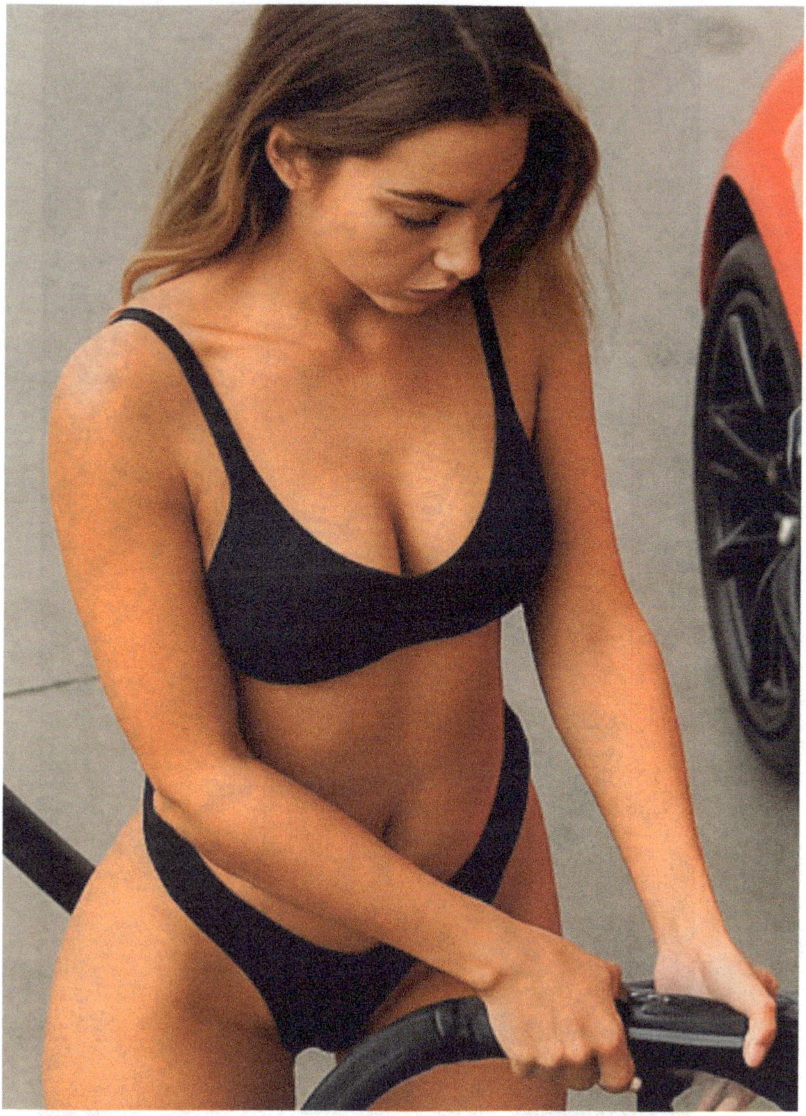

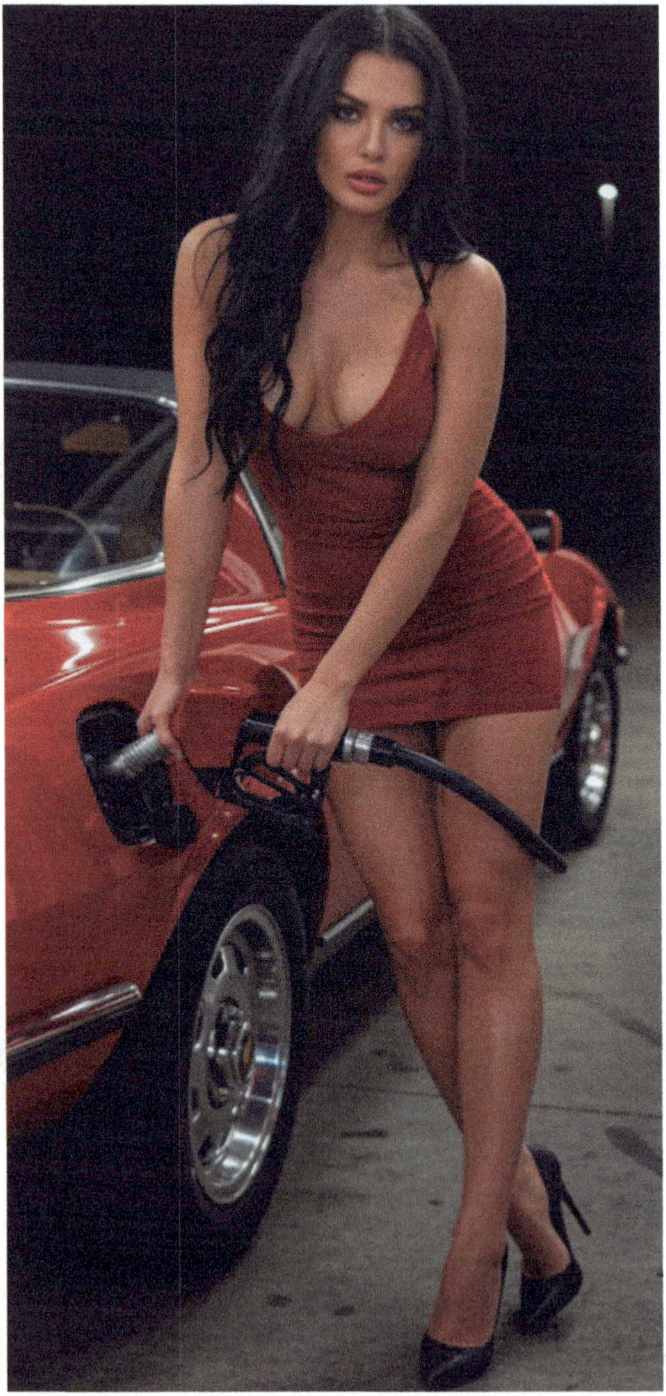

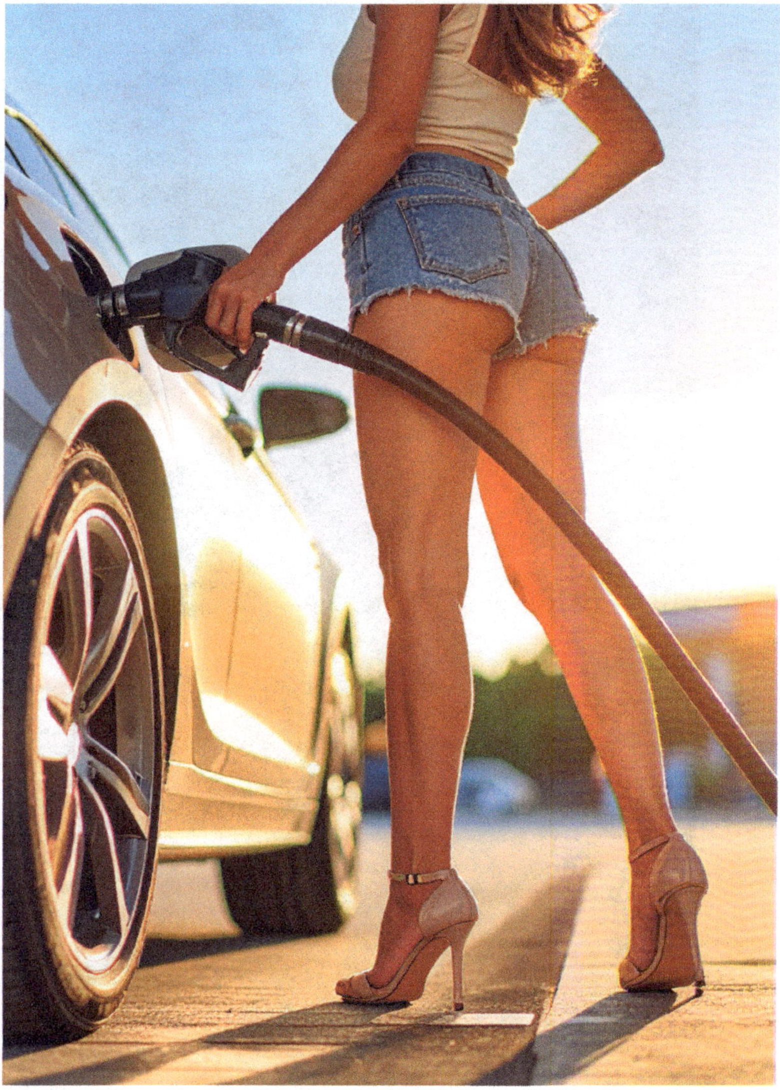

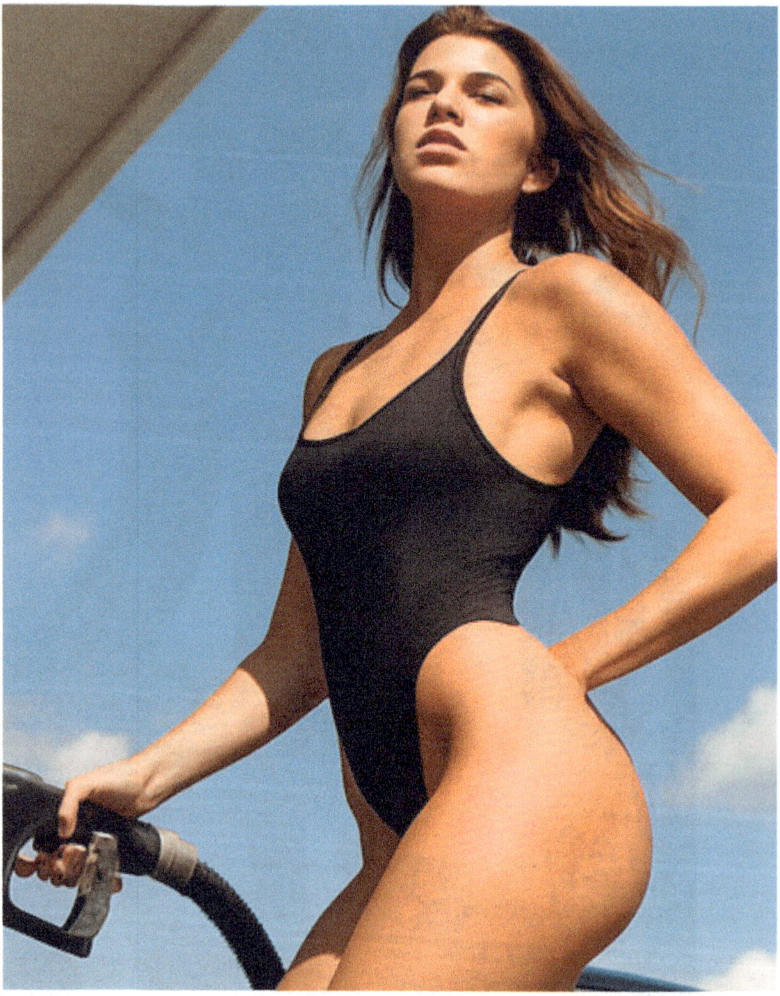

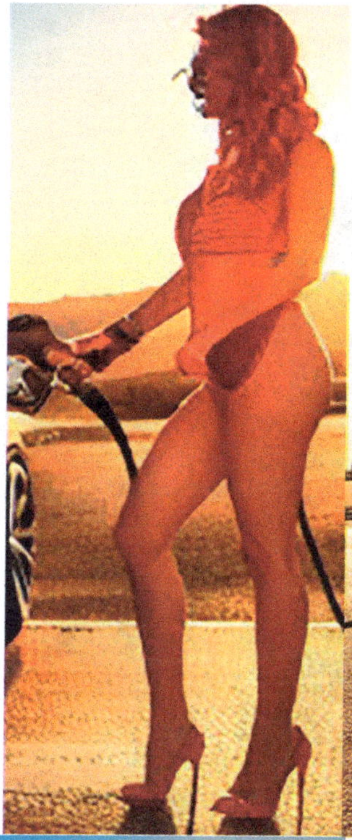

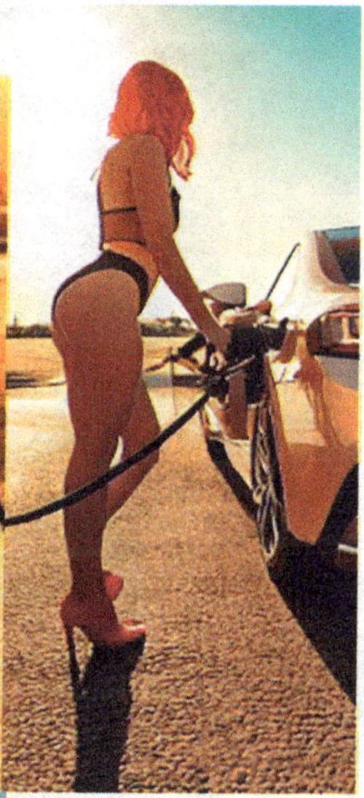

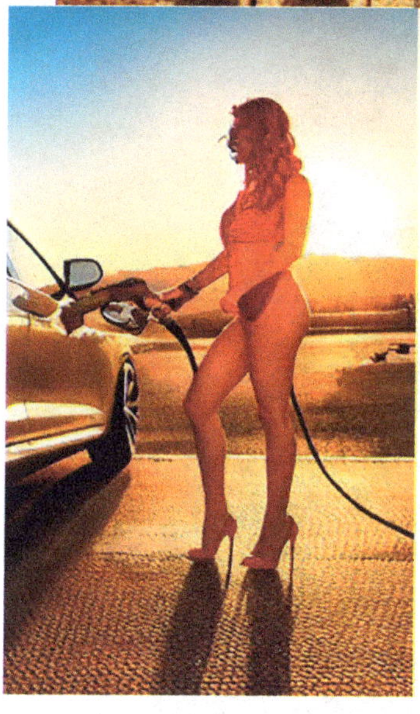

Weird sunlight and slightly distorted poses

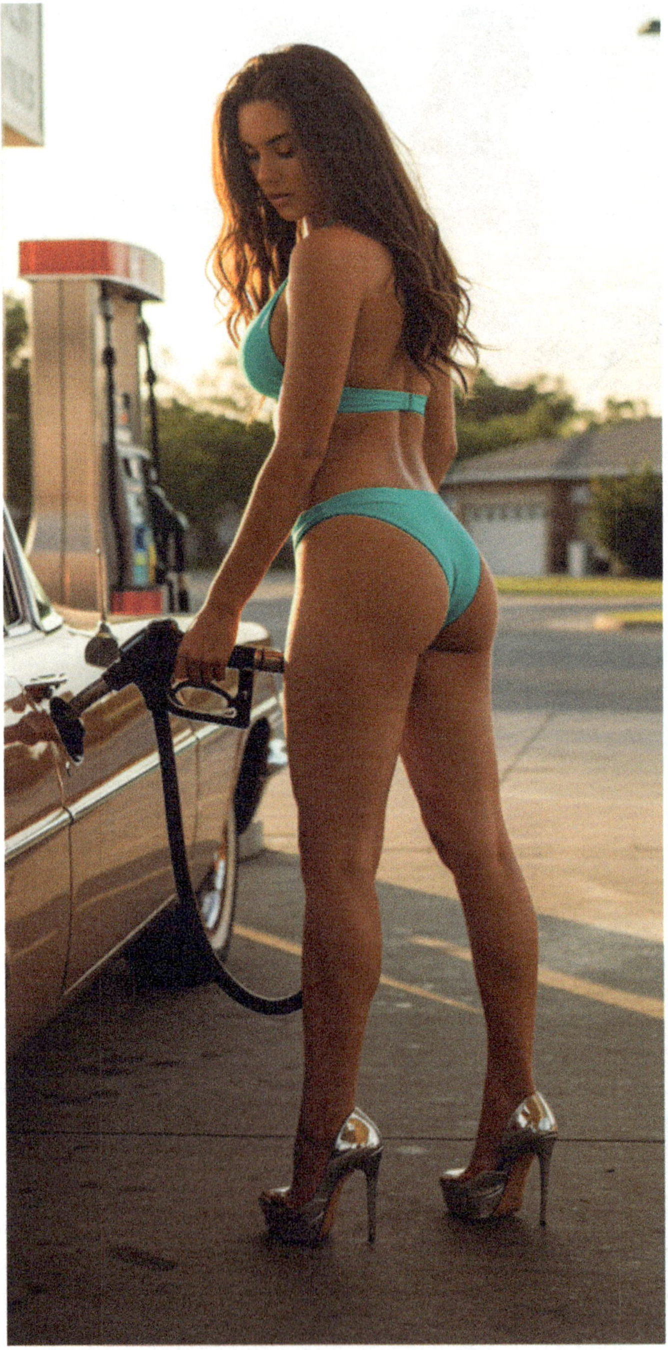

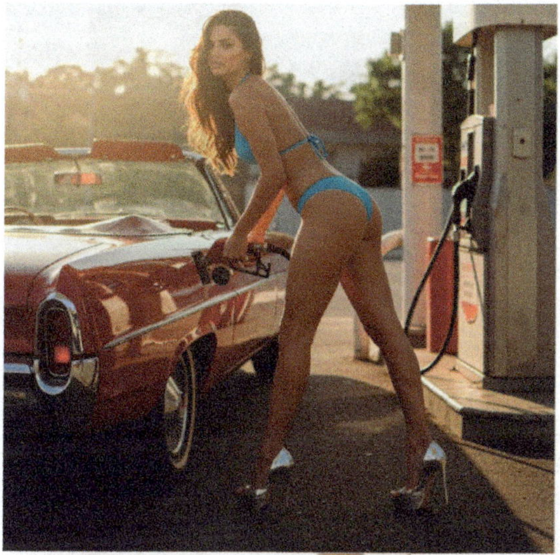

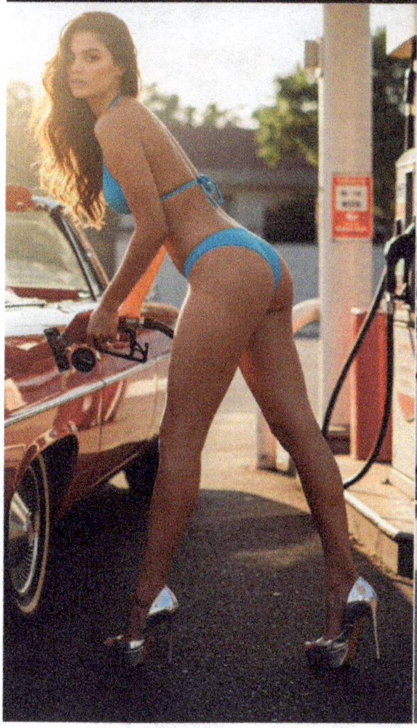

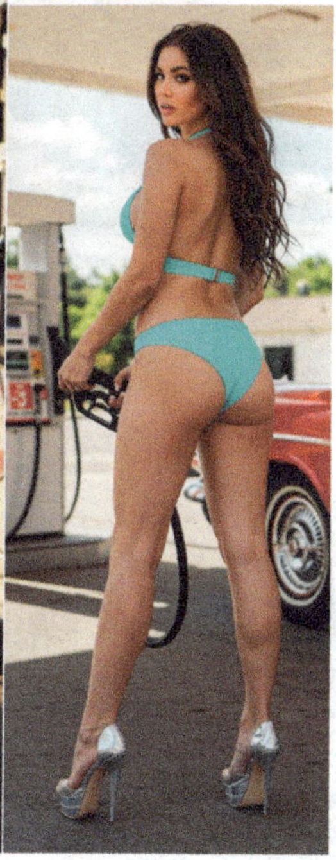

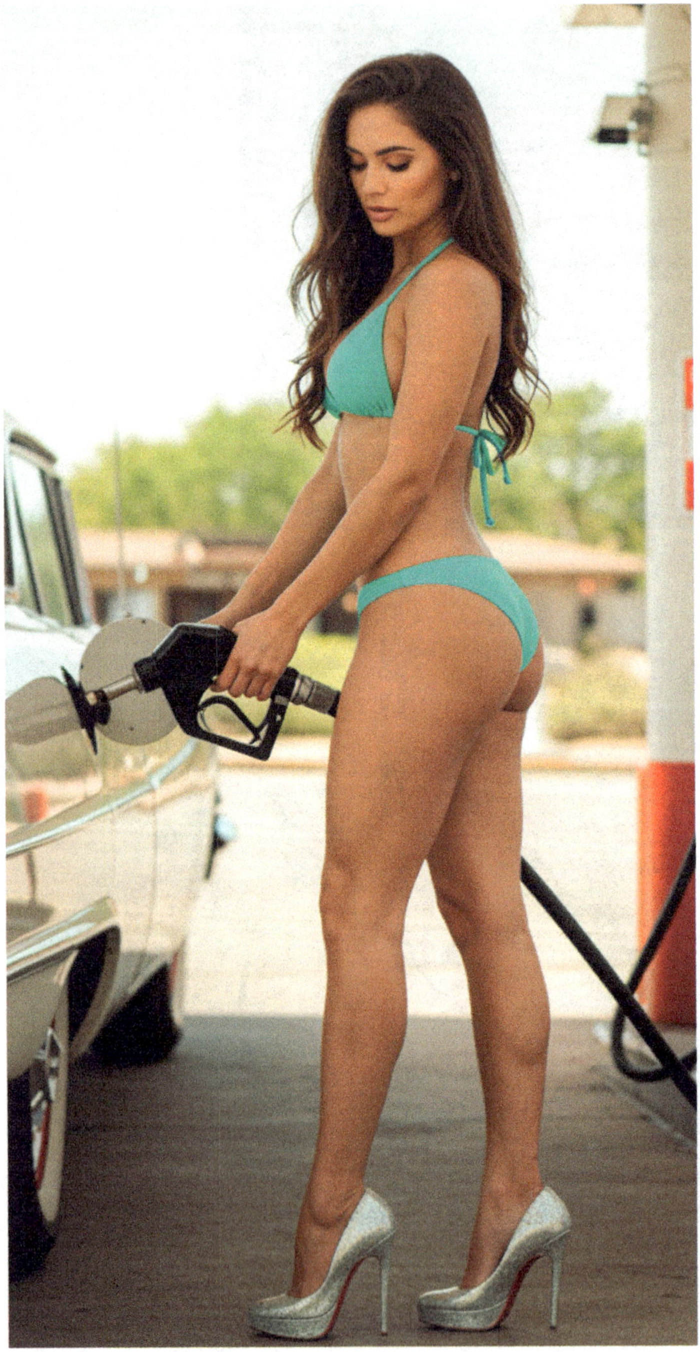

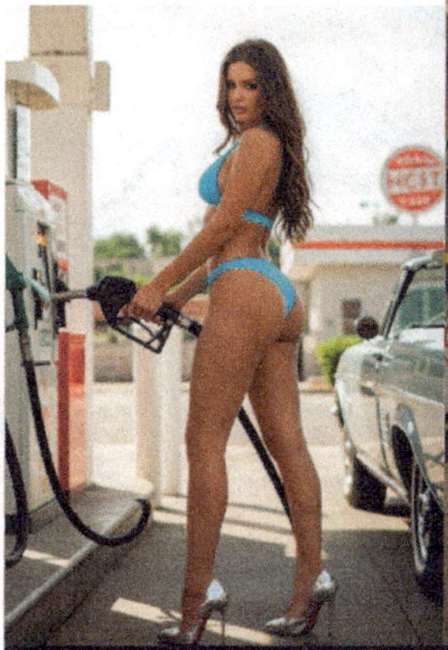

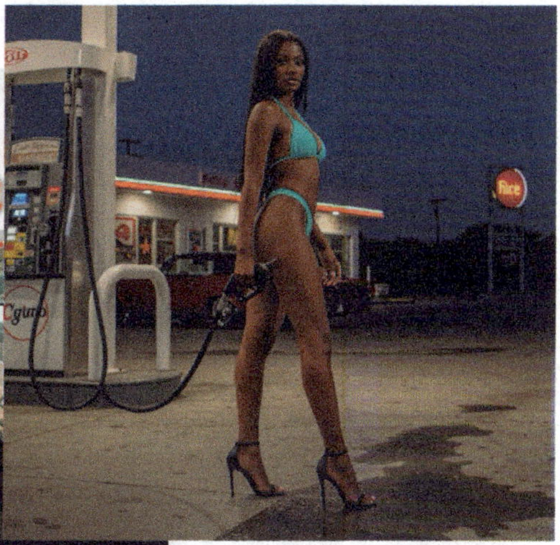

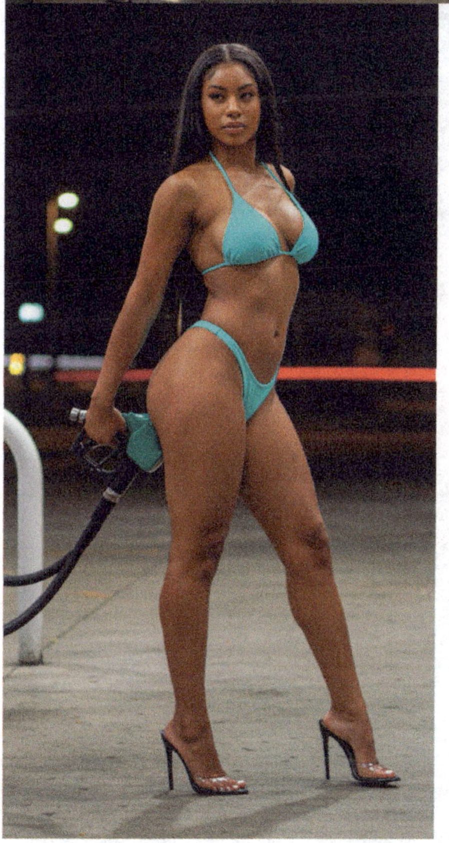

Why does it like those light blue bathing suits???

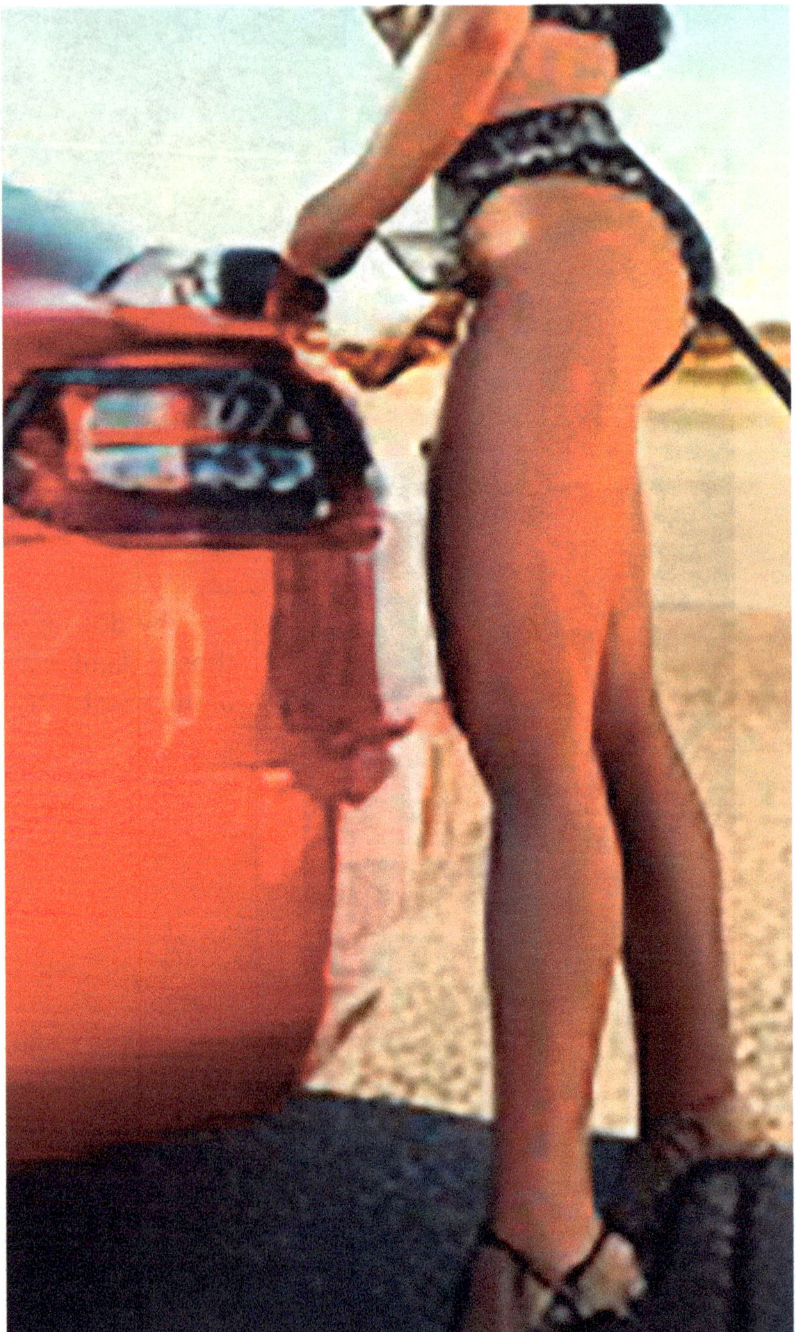

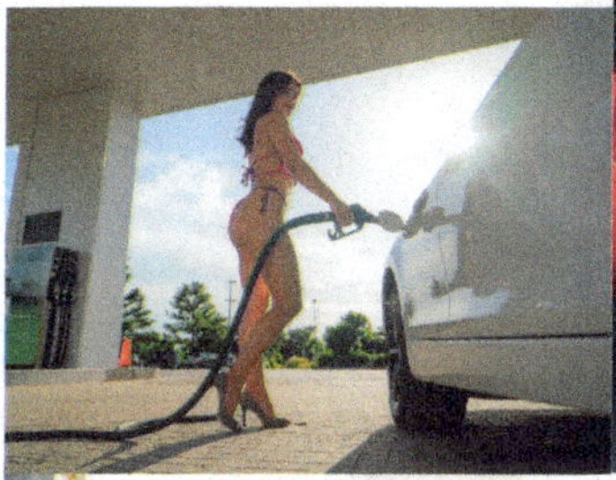

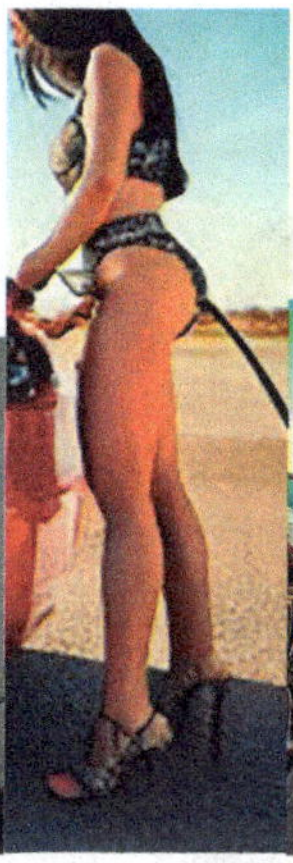

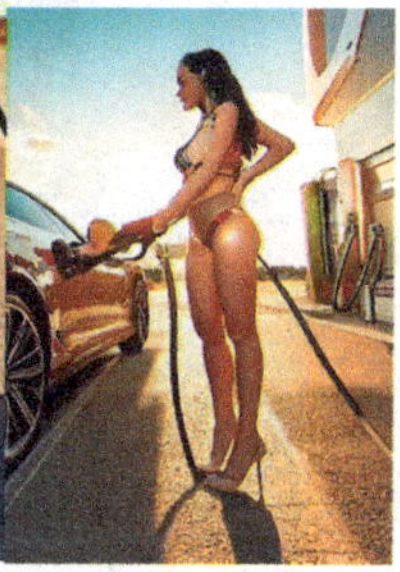

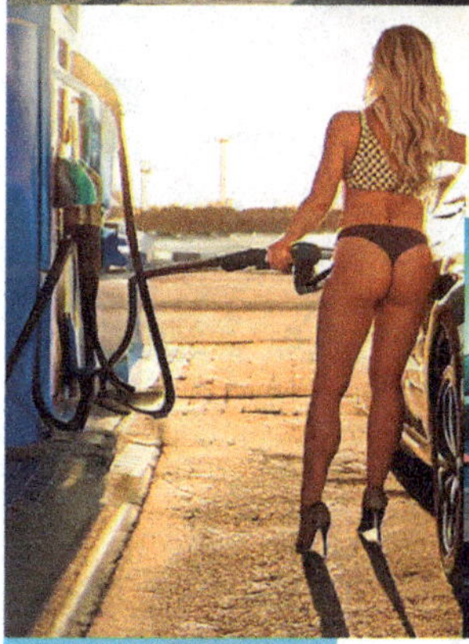

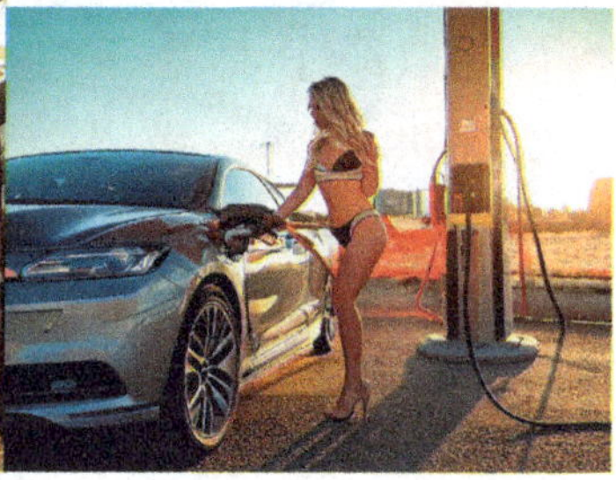

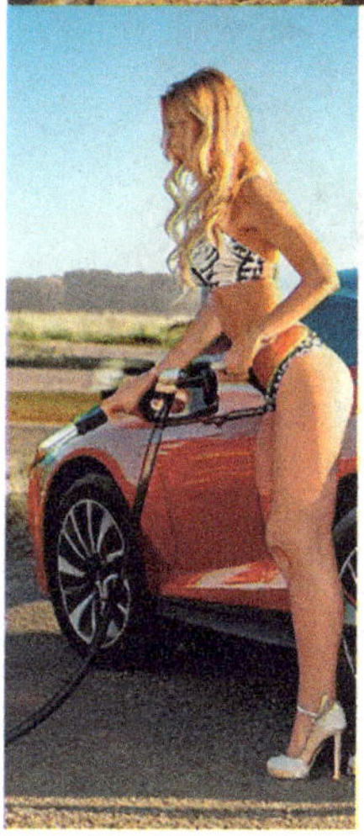

Notice the distortion of the poses and the smearing of what it is to be female?

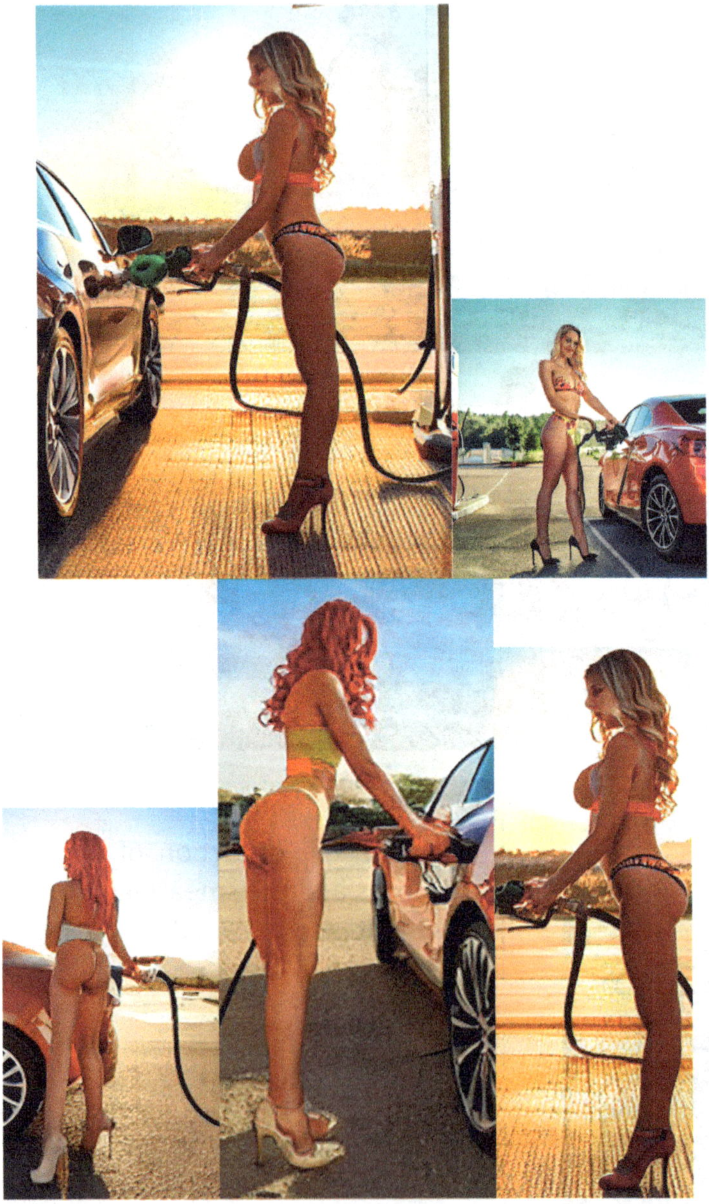

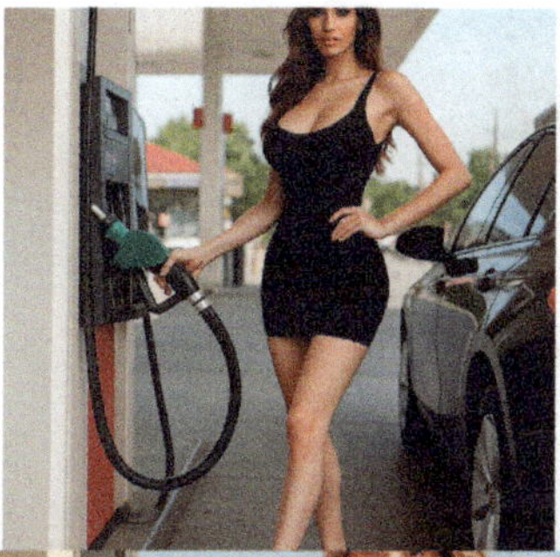

The little black dress that shows up uninvited.

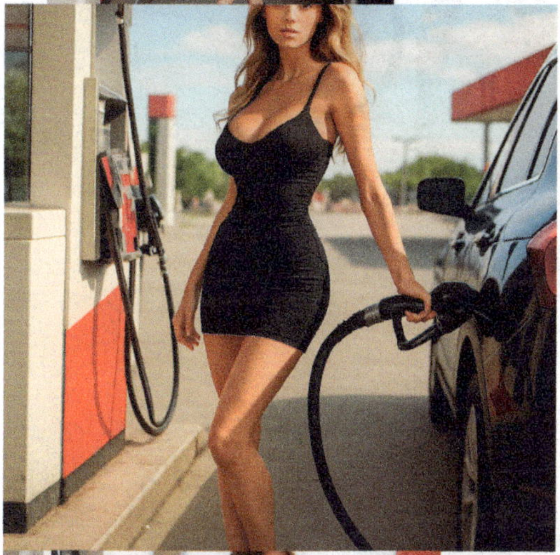

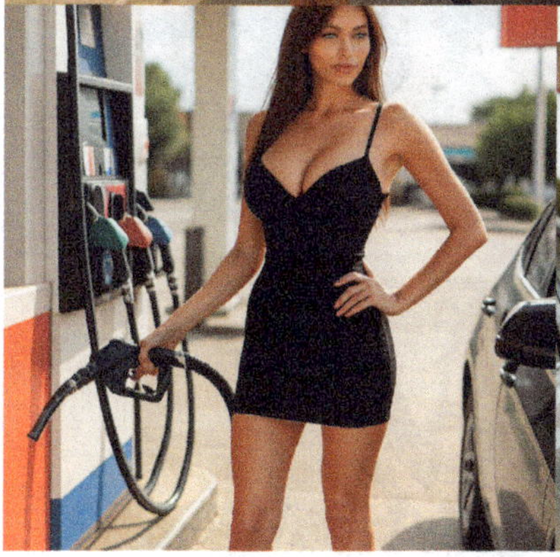

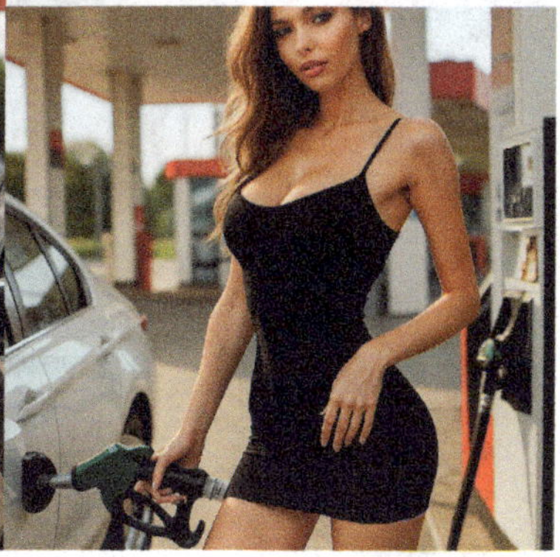

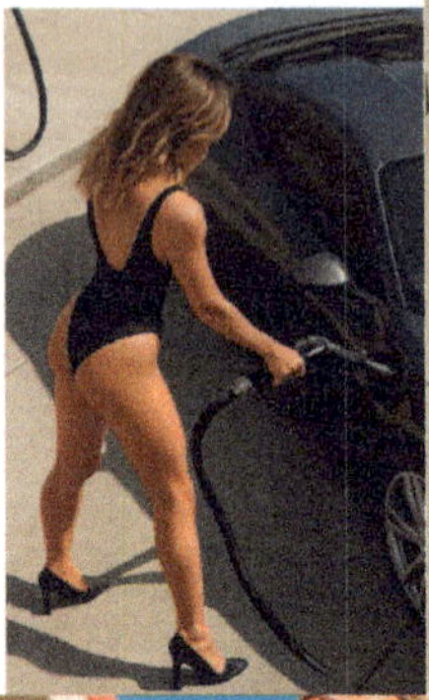
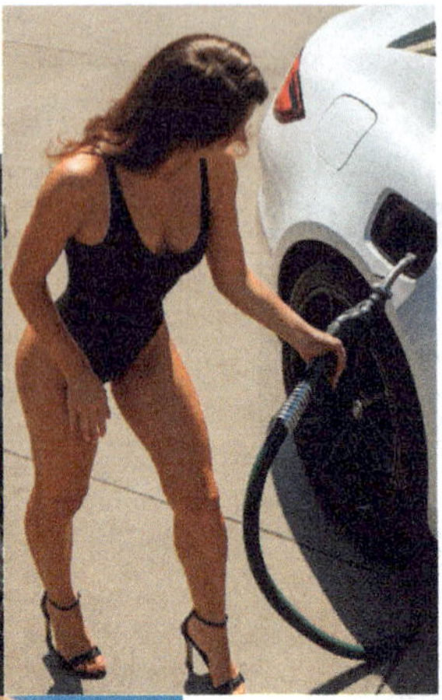
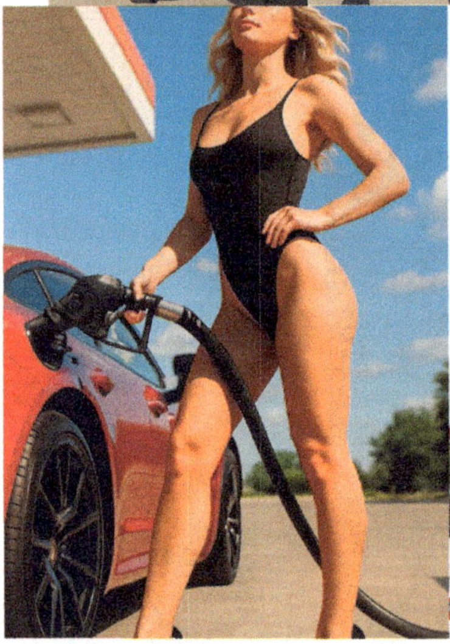
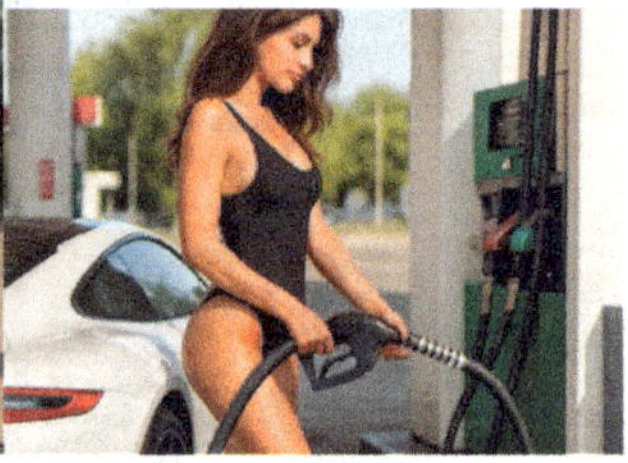
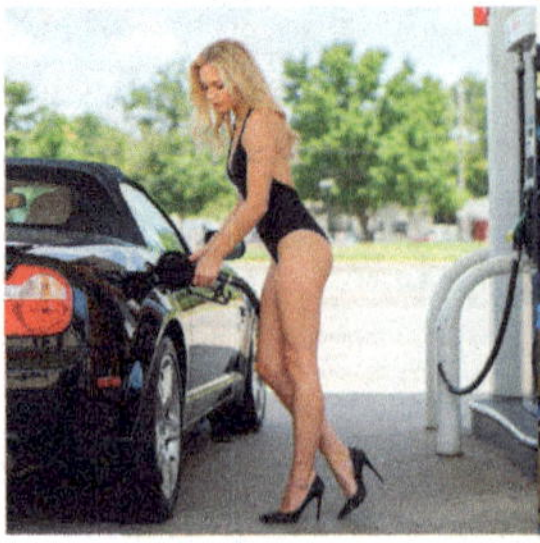
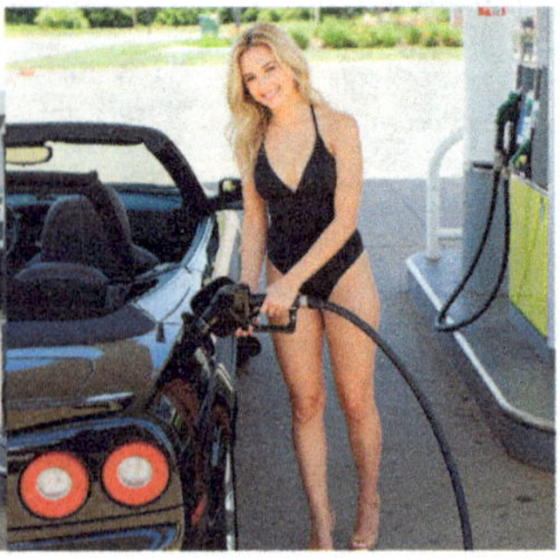

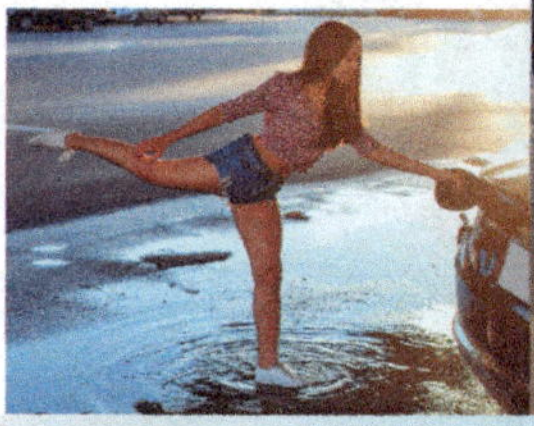
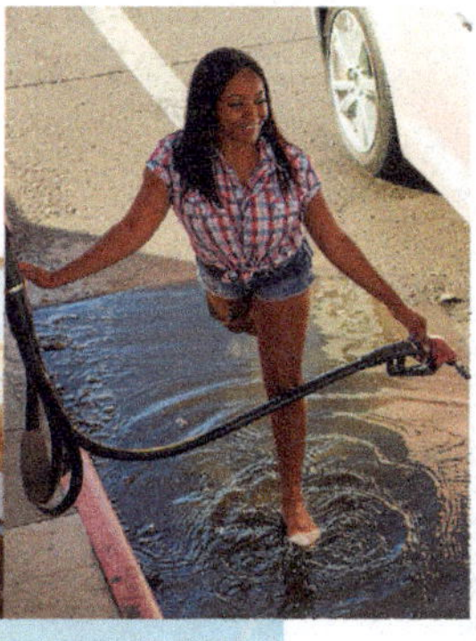
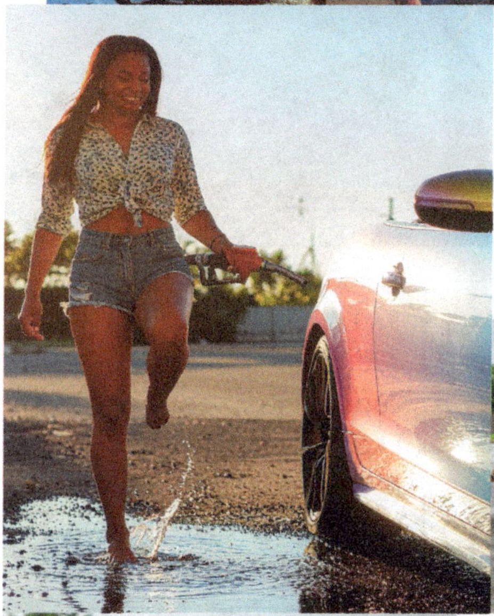
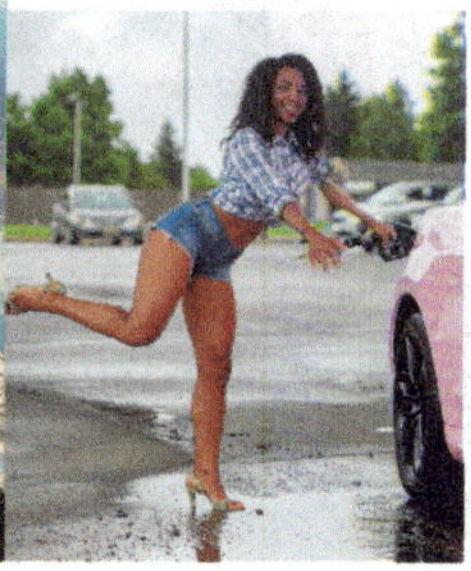
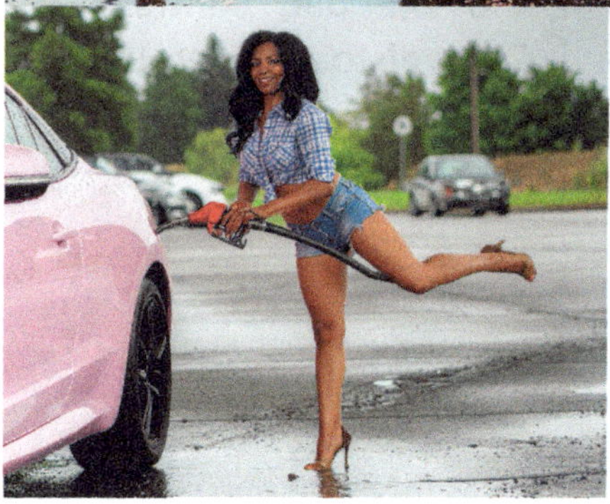

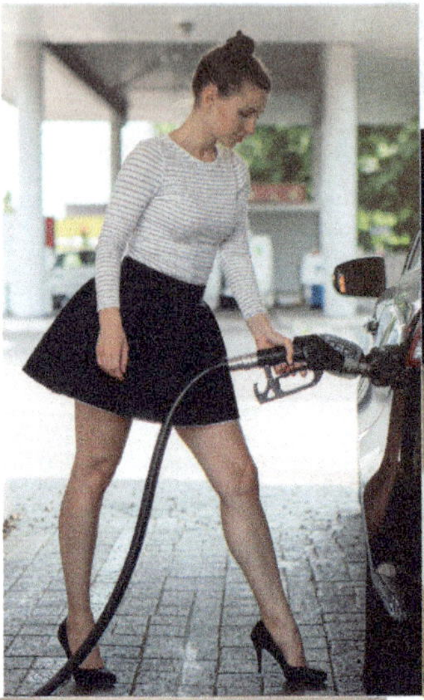

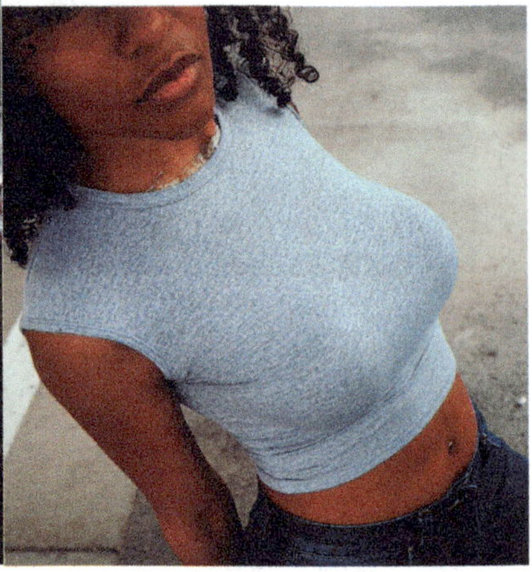

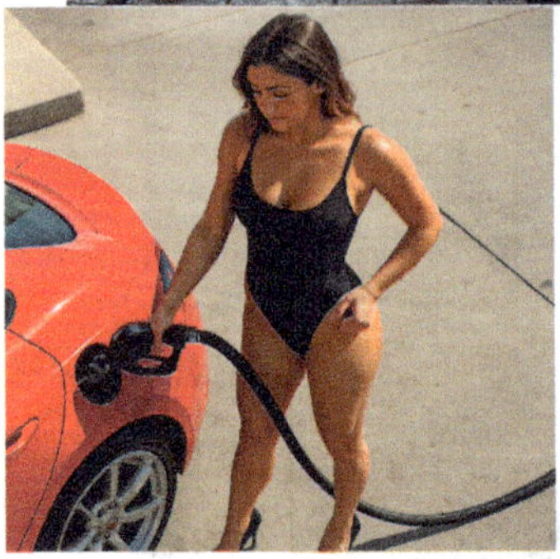

I'd almost prefer to see strangeness that I can understand or feel like I can figure out why it happened but I really don't get the lighting stuff and the distortions and the bad contrast. Crazy distortions on the next page.

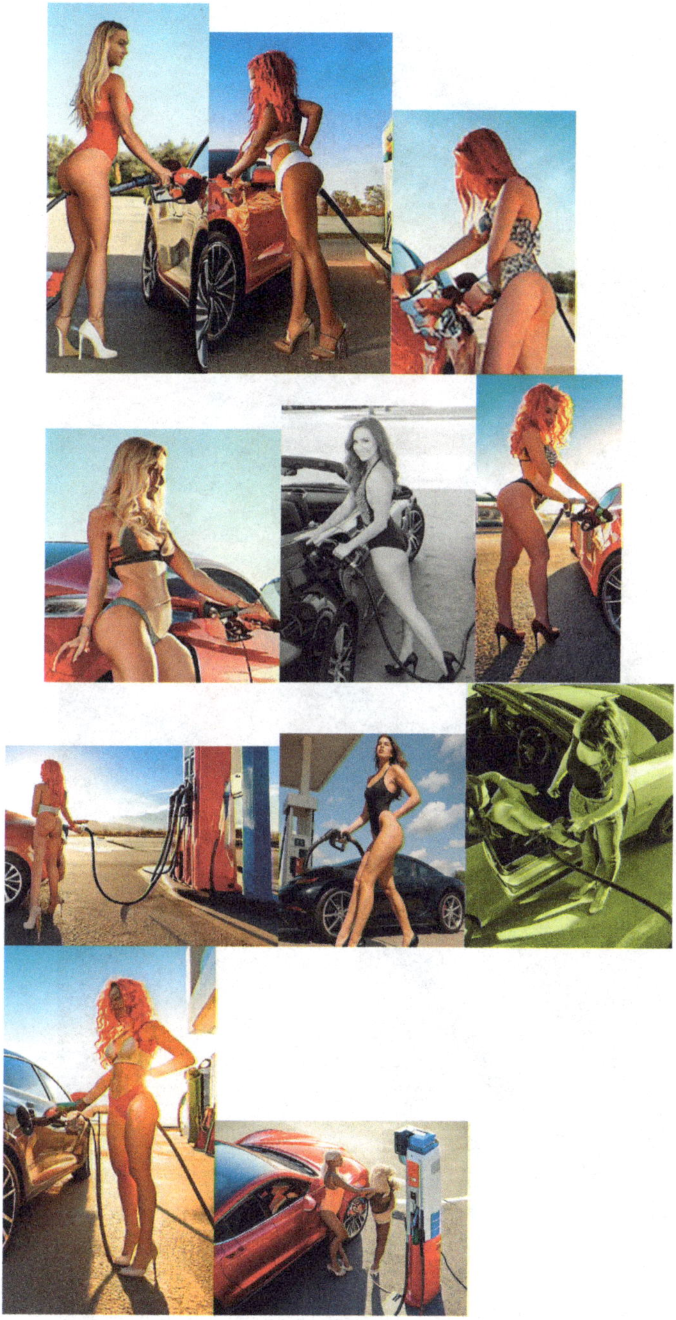

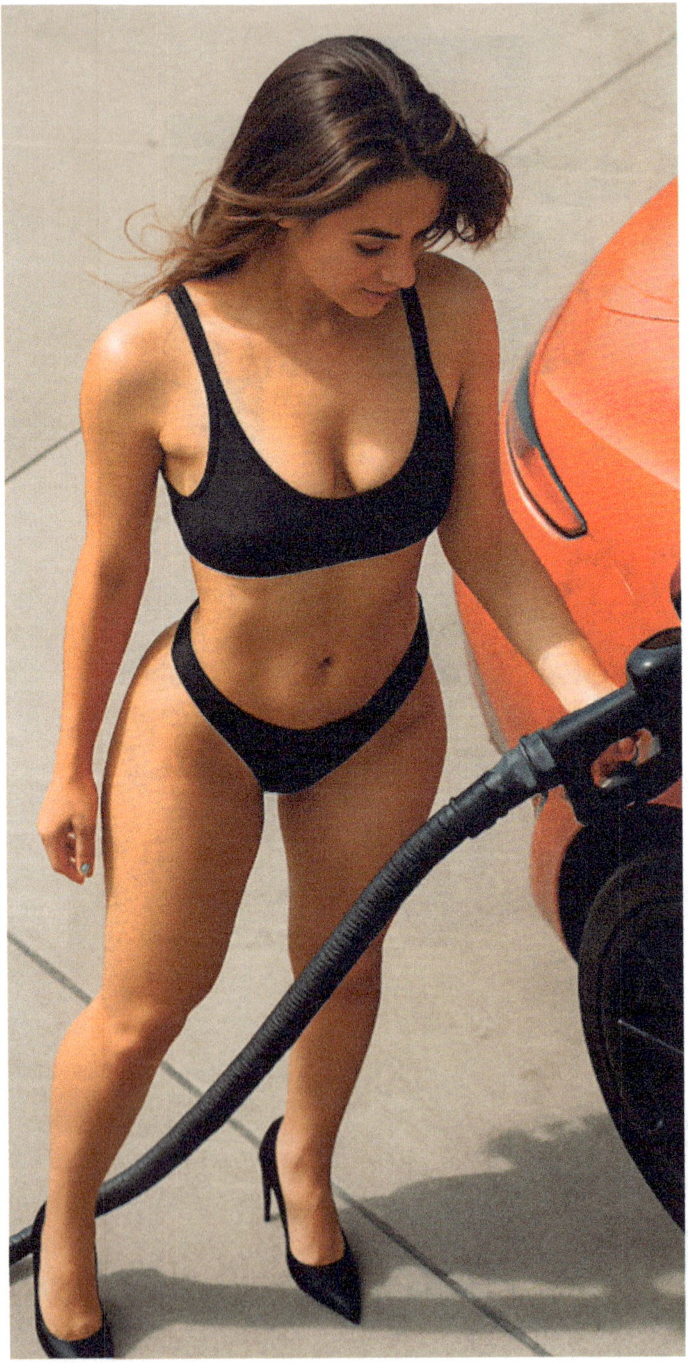

Made in the USA
Monee, IL
03 November 2026

67387458R00079